Wadsworth Classic Readings in Sociology

Wadsworth Classic Readings in Sociology

FIFTH EDITION

Australia • Brazil • Japan • Korea • Mexico • Singapore • Spain • United Kingdom • United States

WADSWORTH
CENGAGE Learning

Wadsworth Classic Readings in Sociology, Fifth Edition

Publisher/Executive Editor: Linda Schreiber-Ganster

Editor: Erin Mitchell

Assistant Editor: Rachael Krapf

Editorial Assistant: Pamela Simon

Media Editor: Melanie Cregger

Marketing Manager: Andrew Keay

Marketing Coordinator: Jillian Myers

Marketing Communications Manager: Laura Localio

Content Project Manager: Matthew Ballantyne

Art Director: Caryl Gorska

Print Buyer: Paula Vang

Permissions Editor: Roberta Broyer

Production Service: PrePressPMG

Cover Designer: RHDG

Cover Image: Getty Images

Compositor: PrePressPMG

For product information and technology assistance, contact us at **Cengage Learning Customer & Sales Support, 1-800-354-9706**

For permission to use material from this text or product, submit all requests online at **www.cengage.com/permissions**. Further permissions questions can be emailed to **permissionrequest@cengage.com**.

Library of Congress Control Number: 2009943634

Student Edition:

ISBN-13: 978-0-495-60276-7

ISBN-10: 0-495-60276-0

Wadsworth
20 Davis Drive
Belmont, CA 94002
USA

Cengage Learning is a leading provider of customized learning solutions with office locations around the globe, including Singapore, the United Kingdom, Australia, Mexico, Brazil, and Japan. Locate your local office at **international.cengage.com/region**.

Cengage Learning products are represented in Canada by Nelson Education, Ltd.

For your course and learning solutions, visit **www.cengage.com.**

Purchase any of our products at your local college store or at our preferred online store **www.CengageBrain.com**.

Printed in the United States of America
1 2 3 4 5 6 7 14 13 12 11 10

Contents

Preface

Classic Readings in Sociology has been carefully designed to include selections that teachers of introductory sociology and theory courses most often want their students to read. The articles were selected to represent past and present sociological thought that has endured and will endure, establishing these selections as classics in the field. The low price of this collection makes it an ideal accompaniment to a standard textbook—or for instructors who don't use a text, this reader (perhaps used with a selection of other books) gives students a strong introduction to the foundation of the field.

The articles were chosen after gathering feedback from several reviewers and reviewing results of an extensive survey of instructors' input of selections they consider classics in their teaching of sociology. Included are excerpts from timeless, well-known works, such as C. Wright Mills's "The Promise of Sociology" and Peter Berger's "Invitation to Sociology" as well as selections from highly regarded contemporary writings, such as Jonathan Kozol's "Savage Inequalities." These works help students develop a sense of the nature of the discipline and its history, exposing them to original thinking and analysis.

The collection includes readings that can accompany each part of an introductory course. For instance, Peter Berger's "Invitation to Sociology" provides a perfect way to introduce the nature of sociology to students. "The Self" by George Herbert Mead can easily supplement a discussion of socialization; an excerpt from Mills's "The Power Elite" can augment work on the political world. The articles can promote stimulating classroom discussions and form the basis for engaging writing assignments. Most important, the articles provide the tools to help students understand how sociologists think, and they lay the foundation for future reading and learning.

Classic Readings in Sociology includes several additional helpful elements. A full glossary of sociological terms will especially benefit instructors who choose to use this collection without an accompanying textbook. Website links are provided to help students and instructors access recent data and information. These links can provide another way to prompt classroom discussion, connect the classic articles to more recent writings, and supplement research for a written paper or exercise. Finally, discussion questions are provided for each article.

What is new in the fifth edition? Previous editions enjoyed resoundingly positive feedback. The selections have been validated as truly seminal works in the field. To update the collection, we continue to add classics and delete the least used articles (as best we can determine) to keep the volume reasonable in length and cost. The newest article is Ida B. Wells-Barnett's "Lynch Law in America." Reviewers in the previous edition specifically requested Émile Durkheim's "The Normality of Crime," Max Weber's "The Protestant Ethic and the Spirit of Capitalism," and "Women and Economics" by Charlotte Perkins Gilman for their classes, and for this edition, we felt the time had come for the collection to include Wells-Barnett and her poignant piece on the brutal lynch laws that were practiced in our society.

Wadsworth Classic Readings in Sociology

The Promise of Sociology

C. WRIGHT MILLS

Nowadays men often feel that their private lives are a series of traps. They sense that within their everyday worlds, they cannot overcome their troubles, and in this feeling, they are often quite correct: What ordinary men are directly aware of and what they try to do are bounded by the private orbits in which they live; their visions and their powers are limited to the close-up scenes of job, family, neighborhood; in other milieux, they move vicariously and remain spectators. And the more aware they become, however vaguely, of ambitions and of threats which transcend their immediate locales, the more trapped they seem to feel.

Underlying this sense of being trapped are seemingly impersonal changes in the very structure of continent-wide societies. The facts of contemporary history are also facts about the success and the failure of individual men and women. When a society is industrialized, a peasant becomes a worker; a feudal lord is liquidated or becomes a businessman. When classes rise or fall, a man is employed or unemployed; when the rate of investment goes up or down, a man takes new heart or goes broke. When wars happen, an insurance salesman becomes a rocket launcher; a store clerk, a radar man; a wife lives alone; a child grows up without a father. Neither the life of an individual nor the history of a society can be understood without understanding both.

Yet men do not usually define the troubles they endure in terms of historical change and institutional contradiction. The well-being they enjoy, they do not usually impute to the big ups and downs of the societies in which they live. Seldom aware of the intricate connection between the patterns of their own lives and the course of world history, ordinary men do not usually know what this connection means for the kinds of men they are becoming and for the kinds of

SOURCE: From *The Sociology Imagination* by C. Wright Mills, 2000 by Oxford University Press, Inc. Used by permission of Oxford University Press, Inc.

history-making in which they might take part. They do not possess the quality of mind essential to grasp the interplay of man and society, or biography and history, of self and world. They cannot cope with their personal troubles in such ways as to control the structural transformations that usually lie behind them.

Surely it is no wonder. In what period have so many men been so totally exposed at so fast a pace to such earthquakes of change? That Americans have not known such catastrophic changes as have the men and women of other societies is due to historical facts that are now quickly becoming "merely history." The history that now affects every man is world history. Within this scene and this period, in the course of a single generation, one-sixth of mankind is transformed from all that is feudal and backward into all that is modern, advanced, and fearful. Political colonies are freed; new and less visible forms of imperialism installed. Revolutions occur; men feel the intimate grip of new kinds of authority. Totalitarian societies rise, and are smashed to bits—or succeed fabulously. After two centuries of ascendancy, capitalism is shown up as only one way to make society into an industrial apparatus. After two centuries of hope, even formal democracy is restricted to a quite small portion of mankind. Everywhere in the underdeveloped world, ancient ways of life are broken up and vague expectations become urgent demands. Everywhere in the overdeveloped world, the means of authority and of violence become total in scope and bureaucratic in form. Humanity itself now lies before us, the supernation at either pole concentrating its most coordinated and massive efforts upon the preparation of World War Three.

The very shaping of history now outpaces the ability of men to orient themselves in accordance with cherished values. And which values? Even when they do not panic, men often sense that older ways of feeling and thinking have collapsed and that newer beginnings are ambiguous to the point of moral stasis. Is it any wonder that ordinary men feel they cannot cope with the larger worlds with which they are so suddenly confronted? That they cannot understand the meaning of their epoch for their own lives? That—in defense of selfhood—they become morally insensible, trying to remain altogether private men? Is it any wonder that they come to be possessed by a sense of the trap?

It is not only information that they need—in this Age of Fact, information often dominates their attention and overwhelms their capacities to assimilate it. It is not only the skills of reason that they need—although their struggles to acquire these often exhaust their limited moral energy.

What they need, and what they feel they need, is a quality of mind that will help them to use information and to develop reason in order to achieve lucid summations of what is going on in the world and of what may be happening within themselves. It is this quality, I am going to contend, that journalists and scholars, artists and publics, scientists and editors are coming to expect of what may be called the sociological imagination.

The sociological imagination enables its possessor to understand the larger historical scene in terms of its meaning for the inner life and the external career of a variety of individuals. It enables him to take into account how individuals, in the welter of their daily experience, often become falsely conscious of their social positions. Within that welter, the framework of modern society is sought, and within that framework the psychologies of a variety of men and women are formulated. By such means the personal uneasiness of individuals is focused upon explicit troubles and the indifference of publics is transformed into involvement with public issues.

The first fruit of this imagination—and the first lesson of the social science that embodies it—is the idea that the individual can understand his own experience and gauge his own fate only by locating himself within his period, that he can know his own chances in life by becoming aware of those of all individuals in his circumstances. In many ways it is a terrible lesson; in many ways a magnificent one. We do not know the limits of man's capacities for supreme effort or willing degradation, for agony or glee, for pleasurable brutality or the sweetness of reason. But in our time we have come to know that the limits of "human nature" are frighteningly broad. We have come to know that every individual lives, from one generation to the next, in some society; that he lives out a biography, and that he lives it out within some historical sequence. By the fact of his living he contributes, however minutely, to the shaping of this society and to the course of its history, even as he is made by society and by its historical push and shove.

The sociological imagination enables us to grasp history and biography and the relations between the two within society. That is its task and its promise. To recognize this task and this promise is the mark of the classic social analyst. It is characteristic of Herbert Spencer—turgid, polysyllabic, comprehensive; of E. A. Ross—graceful, muckraking, upright; of August Comte and Émile Durkheim; of the intricate and subtle Karl Mannheim. It is the quality of all that is intellectually excellent in Karl Marx; it is the clue to Thorstein Veblen's brilliant and ironic insight, to Joseph Schumpeter's many-sided constructions of reality; it is the basis of the psychological sweep of W. E. H. Lecky no less than of the profundity and clarity of Max Weber. And it is the signal of what is best in contemporary studies of man and society.

No social study that does not come back to the problems of biography, of history, and of their intersections within a society has completed its intellectual journey. Whatever the specific problems of the classic social analysts, however limited or however broad the features of social reality they have examined, those who have been imaginatively aware of the promise of their work have consistently asked three sorts of questions:

1. What is the structure of this particular society as a whole? What are its essential components, and how are they related to one another? How does it differ from other varieties of social order? Within it, what is the meaning of any particular feature for its continuance and for its change?

2. Where does this society stand in human history? What are the mechanics by which it is changing? What is its place within and its

meaning for the development of humanity as a whole? How does any particular feature we are examining affect, and how is it affected by, the historical period in which it moves? And this period—what are its essential features? How does it differ from other periods? What are its characteristic ways of history-making?

3. What varieties of men and women now prevail in this society and in this period? And what varieties are coming to prevail? In what ways are they selected and formed, liberated and repressed, made sensitive and blunted? What kinds of "human nature" are revealed in the conduct and character we observe in this society in this period? And what is the meaning for "human nature" of each and every feature of the society we are examining?

Whether the point of interest is a great power state or a minor literary mood, a family, a prison, a creed—these are the kinds of questions the best social analysts have asked. They are the intellectual pivots of classic studies of man in society—and they are questions inevitably raised by any mind possessing the sociological imagination. For that imagination is the capacity to shift from one perspective to another—from the political to the psychological; from examination of a single family to comparative assessment of the national budgets of the world; from the theological school to the military establishment; from considerations of an oil industry to studies of contemporary poetry. It is the capacity to range from the most impersonal and remote transformations to the most intimate features of the human self—and to see the relations between the two. Back of its use there is always the urge to know the social and historical meaning of the individual in the society and in the period in which he has his quality and his being.

That, in brief, is why it is by means of the sociological imagination that men now hope to grasp what is going on in the world, and to understand what is happening in themselves as minute points of the intersections of biography and history within society. In large part, contemporary man's selfconscious view of himself as at least an outsider, if not a permanent stranger, rests upon an absorbed realization of social relativity and of the transformative power of history. The sociological imagination is the most fruitful form of this self-consciousness. By its use men whose mentalities have swept only a series of limited orbits often come to feel as if suddenly awakened in a house with which they had only supposed themselves to be familiar. Correctly or incorrectly, they often come to feel that they can now provide themselves with adequate summations, cohesive assessments, comprehensive orientations. Older decisions that once appeared sound now seem to them products of a mind unaccountably dense. Their capacity for astonishment is made lively again. They acquire a new way of thinking, they experience a transvaluation of values: In a word, by their reflection and by their sensibility, they realize the cultural meaning of the social sciences.

Perhaps the most fruitful distinction with which the sociological imagination works is between "the personal troubles of milieu" and "the public issues of social structure." This distinction is an essential tool of the sociological imagination and a feature of all classic work in social science.

Troubles occur within the character of the individual and within the range of his immediate relations with others; they have to do with his self and with those limited areas of social life of which he is directly and personally aware. Accordingly, the statement and the resolution of troubles properly lie within the individual as a biographical entity and within the scope of this immediate milieu— the social setting that is directly open to his personal experience and to some extent his willful activity. A trouble is a private matter: Values cherished by an individual are felt by him to be threatened.

Issues have to do with matters that transcend these local environments of the individual and the range of his inner life. They have to do with the organization of many such milieux into the institutions of an historical society as a whole, with the ways in which various milieux overlap and interpenetrate to form the larger structure of social and historical life. An issue is a public matter: Some value cherished by publics is felt to be threatened. Often there is a debate about what that value really is and

about what it is that really threatens it. This debate is often without focus if only because it is the very nature of an issue, unlike even widespread trouble, that it cannot very well be defined in terms of the immediate and everyday environments of ordinary men. An issue, in fact, often involves a crisis in institutional arrangements, and often too it involves what Marxists call "contradictions" or "antagonisms."

In these terms, consider unemployment. When, in a city of 100,000, only one man is unemployed, that is his personal trouble, and for its relief we properly look to the character of the man, his skills, and his immediate opportunities. But when in a nation of 50 million employees, 15 million men are unemployed, that is an issue, and we may not hope to find its solution within the range of opportunities open to any one individual. The very structure of opportunities has collapsed. Both the correct statement of the problem and the range of possible solutions require us to consider the economic and political institutions of the society, and not merely the personal situation and character of a scatter of individuals.

Consider war. The personal problem of war, when it occurs, may be how to survive it or how to die in it with honor; how to make money out of it; how to climb into the higher safety of the military apparatus; or how to contribute to the war's termination. In short, according to one's values, to find a set of milieux and within it to survive the war or make one's death in it meaningful. But the structural issues of war have to do with its causes; with what types of men it throws up into command; with its effects upon economic and political, family and religious institutions, with the unorganized irresponsibility of a world of nation-states.

Consider marriage. Inside a marriage a man and a woman may experience personal troubles, but when the divorce rate during the first four years of marriage is 250 out of every 1,000 attempts, this is an indication of a structural issue having to do with the institutions of marriage and the family and other institutions that bear upon them.

Or consider the metropolis—the horrible, beautiful, ugly, magnificent sprawl of the great city. For many upper-class people, the personal solution to "the problem of the city" is to have an apartment with private garage under it in the heart of the city and, forty miles out, a house by Henry Hill, garden by Garrett Eckbo, on a hundred acres of private land. In these two controlled environments—with a small staff at each end and a private helicopter connection—most people could solve many of the problems of personal milieux caused by the facts of the city. But all this, however splendid, does not solve the public issues that the structural fact of the city poses. What should be done with this wonderful monstrosity? Break it up into scattered units, combining residence and work? Refurbish it as it stands? Or, after evacuation, dynamite it and build new cities according to new plans in new places? What should those plans be? And who is to decide and to accomplish whatever choice is made? These are structural issues; to confront them and to solve them requires us to consider political and economic issues that affect innumerable milieu.

Insofar as an economy is so arranged that slumps occur, the problem of unemployment becomes incapable of personal solution. Insofar as war is inherent in the nation-state system and in the uneven industrialization of the world, the ordinary individual in his restricted milieu will be powerless—with or without psychiatric aid—to solve the troubles this system or lack of system imposes upon him. Insofar as the family as an institution turns women into darling little slaves and men into their chief providers and unweaned dependents, the problem of a satisfactory marriage remains incapable of purely private solution. Insofar as the overdeveloped megalopolis and the overdeveloped automobile are built-in features of the overdeveloped society, the issues of urban living will not be solved by personal ingenuity and private wealth.

What we experience in various and specific milieux, I have noted, is often caused by structural changes. Accordingly, to understand the changes of many personal milieux we are required to look beyond them. And the number and variety of such structural changes increase as the institutions within which we live become more embracing and more intricately connected with one another. To be aware of the idea of social structure and to use it with sensibility is to be capable of tracing such linkages among a great variety of milieux. To be able to do that is to possess the sociological imagination....

DISCUSSION QUESTIONS

1. C. Wright Mills argues that personal troubles can be understood in terms of large-scale patterns that extend beyond individual experience and are part of society and history. Identify an issue that you think would be a personal trouble for an individual and then identify the societal influences you think impinge on the individual's experience. (Hint: You might think of such things as divorce, violence, or school failure.)

2. Mills identifies the central task of sociology as grasping the relationships between history and biography. To do this, identify two large-scale historical events that you think have most shaped your biography. Now ask someone of a different generation that same question. What does this tell you about how sociologists think about the relationship between individuals and society?

3. Sociologists sometimes distinguish between micro-level and macro-level theories. Both study social influences on human lives. Take a topic (such as terrorism, divorce, or illness), and using the perspective that Mills articulates, how might someone using a micro-theoretical approach study this topic? How would that differ from how someone using a macro-level theoretical approach would study the same topic?

INTERNET RESOURCES

Suggested Web URLs for Further Study

http://www.ac.wwu.edu/~stephan/timeline.html
This website contains a basic timeline of important people and papers in sociology beginning from its basis in philosophy.

http://www.socioweb.com
This site is an independent guide to sociological resources on the web.

Virtual Society

For further information on this subject, including links to relevant websites, go to the Wadsworth Sociology homepage at *http://www.cengage.com/sociology*.

REFERENCES

Linton, R. 1936. *The study of man*. New York: Appleton-Century.

Malinowski, B. 1948. *Magic, science and religion*. Glencoe, IL: Free Press.

Murdock, G. P. 1949. *Social structure*. New York: Macmillan.

2

Invitation to Sociology

PETER L. BERGER

It can be said that the first wisdom of sociology is this—things are not what they seem. This too is a deceptively simple statement. It ceases to be simple after a while. Social reality turns out to have many layers of meaning. The discovery of each new layer changes the perception of the whole.

Anthropologists use the term "culture shock" to describe the impact of a totally new culture upon a newcomer. In an extreme instance such shock will be experienced by the Western explorer who is told, halfway through dinner, that he is eating the nice old lady he had been chatting with the previous day—a shock with predictable physiological if not moral consequences. Most explorers no longer encounter cannibalism in their travels today. However, the first encounters with polygamy or with puberty rites or even with the way some nations drive their automobiles can be quite a shock to an American visitor. With the shock may go not only disapproval or disgust but a sense of excitement that things can really be that different from what they are at home. To some extent, at least, this is the excitement of any first travel abroad. The experience of sociological discovery could be described as "culture shock" minus geographical displacement. In other words, the sociologist travels at home—with shocking results. He is unlikely to find that he is eating a nice old lady for dinner. But the discovery, for instance, that his own church has considerable money invested in the missile industry or that a few blocks from his home there are people who engage in cultic orgies may not be drastically different in emotional impact. Yet we would not want to imply that sociological discoveries are always or even usually outrageous to moral sentiment. Not at all. What they have in common with exploration in distant lands, however, is the sudden illumination of new and unsuspected facets of human existence in society. This is the excitement and, as we shall try to show later, the humanistic justification of sociology.

SOURCE: From *An Invitation to Sociology* by Peter Berger, 1963. Used by permission of Doubleday, a division of Random House, Inc.

People who like to avoid shocking discoveries, who prefer to believe that society is just what they were taught in Sunday school, who like the safety of the rules and the maxims of what Alfred Schuetz has called the "world-taken-for-granted," should stay away from sociology. People who feel no temptation before closed doors, who have no curiosity about human beings, who are content to admire scenery without wondering about the people who live in those houses on the other side of that river, should probably also stay away from sociology. They will find it unpleasant or, at any rate, unrewarding. People who are interested in human beings only if they can change, convert, or reform them should also be warned, for they will find sociology much less useful than they hoped. And people whose interest is mainly in their own conceptual constructions will do just as well to turn to the study of little white mice. Sociology will be satisfying, in the long run, only to those who can think of nothing more entrancing than to watch men and to understand things human....

To ask sociological questions, then, presupposes that one is interested in looking some distance beyond the commonly accepted or officially defined goals of human actions. It presupposes a certain awareness that human events have different levels of meaning, some of which are hidden from the consciousness of everyday life. It may even presuppose a measure of suspicion about the way in which human events are officially interpreted by the authorities, be they political, juridical, or religious in character. If one is willing to go as far as that, it would seem evident that not all historical circumstances are equally favorable for the development of sociological perspective.

It would appear plausible, in consequence, that sociological thought would have the best chance to develop in historical circumstances marked by severe jolts to the self-conception, especially the official and authoritative and generally accepted self-conception of a culture. It is only in such circumstances that perceptive men are likely to be motivated to think beyond the assertions of this self-conception and, as a result, question the authorities....

Sociological perspective can then be understood in terms of such phrases as "seeing through," "looking behind," very much as such phrases would be employed in common speech—"seeing through his game," "looking behind the scenes"—in other words, "being up on all the tricks."...

We could think of this in terms of a common experience of people living in large cities. One of the fascinations of a large city is the immense variety of human activities taking place behind the seemingly anonymous and endlessly undifferentiated rows of houses. A person who lives in such a city will time and again experience surprise or even shock as he discovers the strange pursuits that some men engage in quite unobtrusively in houses that, from the outside, look like all the others on a certain street. Having had this experience once or twice, one will repeatedly find oneself walking down a street, perhaps late in the evening, and wondering what may be going on under the bright lights showing through a line of drawn curtains. An ordinary family engaged in pleasant talk with guests? A scene of desperation amid illness or death? Or a scene of debauched pleasures? Perhaps a strange cult or a dangerous conspiracy? The facades of the houses cannot tell us, proclaiming nothing but an architectural conformity to the tastes of some group or class that may not even inhabit the street any longer. The social mysteries lie behind the facades. The wish to penetrate to these mysteries is an analogon to sociological curiosity. In some cities that are suddenly struck by calamity this wish may be abruptly realized. Those who have experienced wartime bombings know of the sudden encounters with unsuspected (and sometimes unimaginable) fellow tenants in the air-raid shelter of one's apartment building. Or they can recollect the startling morning sight of a house hit by a bomb during the night, neatly sliced in half, the facade torn away and the previously hidden interior mercilessly revealed in the daylight. But in most cities that one may normally live in, the facades must be penetrated by one's own inquisitive intrusions. Similarly, there are historical situations in which the facades of society are violently torn apart and all but the most incurious are forced to see that there was a reality behind the facades all along. Usually this does not happen, and the facades continue to confront us with seemingly rocklike permanence. The perception of the reality behind the facades then demands a considerable intellectual effort.

A few examples of the way in which sociology "looks behind" the facades of social structures might serve to make our argument clearer. Take, for instance, the political organization of a community. If one wants to find out how a modern American city is governed, it is very easy to get the official information about this subject. The city will have a charter, operating under the laws of the state. With some advice from informed individuals, one may look up various statutes that define the constitution of the city. Thus one may find out that this particular community has a city-manager form of administration, or that party affiliations do not appear on the ballot in municipal elections, or that the city government participates in a regional water district. In similar fashion, with the help of some newspaper reading, one may find out the officially recognized political problems of the community.

One may read that the city plans to annex a certain suburban area, or that there has been a change in the zoning ordinances to facilitate industrial development in another area, or even that one of the members of the city council has been accused of using his office for personal gain. All such matters still occur on the, as it were, visible, official, or public level of political life. However, it would be an exceedingly naive person who would believe that this kind of information gives him a rounded picture of the political reality of that community. The sociologist will want to know above all the constituency of the "informal power structure" (as it has been called by Floyd Hunter, an American sociologist interested in such studies), which is a configuration of men and their power that cannot be found in any statutes, and probably cannot be read about in the newspapers. The political scientist or the legal expert might find it very interesting to compare the city charter with the constitutions of other similar communities. The sociologist will be far more concerned with discovering the way in which powerful vested interests influence or even control the actions of officials elected under the charter. These vested interests will not be found

in city hall, but rather in the executive suites of corporations that may not even be located in that community, in the private mansions of a handful of powerful men, perhaps in the offices of certain labor unions, or even, in some instances, in the headquarters of criminal organizations. When the sociologist concerns himself with power, he will "look behind" the official mechanisms that are supposed to regulate power in the community. This does not necessarily mean that he will regard the official mechanisms as totally ineffective or their legal definition as totally illusionary. But at the very least he will insist that there is another level of reality to be investigated in the particular system of power. In some cases he might conclude that to look for real power in the publicly recognized places is quite delusional....

Let us take one further example. In Western countries, and especially in America, it is assumed that men and women marry because they are in love. There is a broadly based popular mythology about the character of love as a violent, irresistible emotion that strikes where it will, a mystery that is the goal of most young people and often of the not-so-young as well. As soon as one investigates, however, which people actually marry each other, one finds that the lightning-shaft of Cupid seems to be guided rather strongly within very definite channels of class, income, education, [and] racial and religious background. If one then investigates a little further into the behavior that is engaged in prior to marriage under the rather misleading euphemism of "courtship," one finds channels of interaction that are often rigid to the point of ritual. The suspicion begins to dawn on one that, most of the time, it is not so much the emotion of love that creates a certain kind of relationship, but that carefully predefined and often planned relationships eventually generate the desired emotion. In other words, when certain conditions are met or have been constructed, one allows oneself "to fall in love." The sociologist investigating our patterns of "courtship" and marriage soon discovers a complex web of motives related in many ways to the entire institutional structure within which an individual lives his life—class, career, economic ambition, aspirations of power and prestige. The miracle of love now begins to look somewhat synthetic. Again, this need not mean in any given instance that the sociologist will declare the romantic interpretation to be an illusion. But, once more, he will look beyond the immediately given and publicly approved interpretations....

We would contend, then, that there is a debunking motif inherent in sociological consciousness. The sociologist will be driven time and again, by the very logic of his discipline, to debunk the social systems he is studying. This unmasking tendency need not necessarily be due to the sociologist's temperament or inclinations. Indeed, it may happen that the sociologist, who as an individual may be of a conciliatory disposition and quite disinclined to disturb the comfortable assumptions on which he rests his own social existence, is nevertheless compelled by what he is doing to fly in the face of what those around him take for granted. In other words, we would contend that the roots of the debunking motif in sociology are not psychological but methodological. The sociological frame of reference, with its built-in procedure of looking for levels of reality other than those given in the official interpretations of society, carries with it a logical imperative to unmask the pretensions and the propaganda by which men cloak their actions with each other. This unmasking imperative is one of the characteristics of sociology particularly at home in the temper of the modern era....

DISCUSSION QUESTIONS

1. Peter L. Berger's argument makes the case for sociology as a debunking perspective—that is, looking behind the taken-for-granted ways that people usually see things. He uses the example of marriage by showing that although people think of marriage as happening just because people are in love, there are actually quite specific social patterns about who marries whom. What social factors—not just romantic factors—do you think influence people's choices of marriage partners?

2. Berger suggests that some historical conditions may be more conducive than others to producing a sociological perspective—particularly if what he calls the "facades of society" are torn apart. Think of an example of such an event, and discuss how this event may have challenged people's prior ways of thinking about society.

3. Implicit in Berger's argument is the idea that those who are content with the status quo may not be as able to acquire a sociological perspective. Why would this be true?

INTERNET RESOURCES

Suggested Web URLs for Further Study

http://www.aacsnet.org
The Society for Applied Sociology, founded in 1978, is an international organization for professionals involved in applying sociological knowledge in a wide variety of settings.

http://www.angelfire.com/or/sociologyshop/PLB.html
This website is dedicated to Peter L. Berger and his work, especially *An Invitation to Sociology* and *The Social Construction of Reality*.

Virtual Society

For further information on this subject, including links to relevant websites, go to the Wadsworth Sociology homepage at *http://www.cengage.com/sociology*.

3

Manifesto of the Communist Party

KARL MARX AND FRIEDRICH ENGELS

BOURGEOIS AND PROLETARIANS[1]

The history of all hitherto existing society[2] is the history of class struggles.

Freeman and slave, patrician and plebeian, lord and self, guild master[3] and journeyman, in a word, oppressor and oppressed, stood in constant opposition to one another, carried on an uninterrupted, now hidden, now open fight, a fight that each time ended, either in a revolutionary reconstitution of society at large, or in the common ruin of the contending classes.

In the earlier epochs of history, we find almost everywhere a complicated arrangement of society into various orders, a manifold gradation of social rank. In ancient Rome we have patricians, knights, plebeians, slaves; in the Middle Ages, feudal lords, vassals, guild-masters, journeymen, apprentices, serfs; in almost all of these classes, again, subordinate gradations.

The modern bourgeois society that has sprouted from the ruins of feudal society has not done away with class antagonisms. It has but established new classes, new conditions of oppression, new forms of struggle in place of the old ones.

Our epoch, the epoch of the bourgeoisie, possesses, however, this distinctive feature; it has simplified the class antagonisms. Society as a whole is more and more splitting up into two great hostile camps, into two great classes directly facing each other: Bourgeoisie and Proletariat.

From the serfs of the Middle Ages sprang the chartered burghers of the earliest towns. From these burgesses the first elements of the bourgeoisie were developed.

The discovery of America, the rounding of the Cape, opened up fresh ground for the rising bourgeoisie. The East Indian and Chinese markets, the [colonization] of America, trade with the colonies, the increase in the means of exchange and in commodities generally, gave to commerce, to navigation, to industry, an impulse never before known, and thereby, to the revolutionary element in the tottering feudal society, a rapid development.

The feudal system of industry, under which industrial production was monopolized by close guilds, now no longer sufficed for the growing wants of the new markets. The manufacturing system took its place. The guild-masters were pushed on one side by the manufacturing middle class; division of labor between the different corporate guilds vanished in the face of division of labor in each single workshop.

Meantime the markets kept ever growing, the demand, ever rising. Even manufacture no longer sufficed. Thereupon, steam and machinery revolutionized industrial production. The place of manufacture was taken by the giant, Modern Industry, the place of the industrial middle class, by industrial millionaires, the leaders of whole industrial armies, the modern bourgeois.

Modern industry has established the world market, for which the discovery of America paved the way. This market has given an immense development to commerce, to navigation, to communication by land. This development has, in its turn, reacted on the extension of industry; and in proportion as industry, commerce, navigation, railways extended, in the same proportion the bourgeoisie developed, increased its capital, and pushed into the background every class handed down from the Middle Ages.

We see, therefore, how the modern bourgeoisie is itself the product of a long course of development, of a series of revolutions in the modes of production and of exchange.

Each step in the development of the bourgeoisie was accompanied by a corresponding political advance of that class. An oppressed class under the sway of the feudal nobility, an armed and self-governing association in the mediaeval commune,[4] here independent urban republic (as in Italy and Germany), there taxable "third estate" of the monarchy (as in France), afterwards, in the period of manufacture proper, serving either the semifeudal or the absolute monarchy as a counterpoise against the nobility, and, in fact, cornerstone of the great

monarchies in general, the bourgeoisie has at last, since the establishment of Modern Industry and of the world-market, conquered for itself, in the modern representative State, exclusive political sway. The executive of the modern State is but a committee for managing the common affairs of the whole bourgeoisie.

The bourgeoisie, historically, has played a most revolutionary part.

The bourgeoisie, wherever it has got the upper hand, has put an end to all feudal, patriarchal, idyllic relations. It has pitilessly torn asunder the motley feudal ties that bound man to his "natural superiors," and has left remaining no other nexus between man and man than naked self-interest, than callous "cash payment." It has drowned the most heavenly ecstasies of religious fervour, of chivalrous enthusiasm, of philistine sentimentalism, in the icy water of egotistical calculation. It has resolved personal worth into exchange value, and in place of the numberless indefeasible chartered freedoms, has set up that single, unconscionable freedom— Free Trade. In one word, for exploitation, veiled by religious and political illusions, it has substituted naked, shameless, direct, brutal exploitation.

The bourgeoisie has stripped of its halo every occupation hitherto honoured and looked up to with reverent awe. It has converted the physician, the lawyer, the priest, the poet, the man of science, into its paid [wage-laborers].

The bourgeoisie has torn away from the family its sentimental veil, and has reduced the family relation to a mere money relation.

The bourgeoisie has disclosed how it came to pass that the brutal display of vigour in the Middle Ages, which Reactionists so much admire, found its fitting complement in the most slothful indolence. It has been the first to show what man's activity can bring about. It has accomplished wonders far surpassing Egyptian pyramids, Roman aqueducts, and Gothic cathedrals; it has conducted expeditions that put in the shade all former Exoduses of nations and crusades.

The bourgeoisie cannot exist without constantly revolutionizing the instruments of production, and thereby the relations of production, and with them the whole relations of society. Conservation of the old modes of production in unaltered form, was, on the contrary, the first condition of existence for all earlier industrial classes. Constant revolutionizing of production, uninterrupted disturbance of all social conditions, everlasting uncertainty and agitation distinguish the bourgeois epoch from all earlier ones. All fixed, fast-frozen relations, which their train of ancient and venerable prejudices and opinions, are swept away, all new formed ones become antiquated before they can ossify. All that is solid melts into air, all that is holy is profaned, and man is at last compelled to face with sober senses, his real conditions of life, and his relations with his kind.

The need of a constantly expanding market for its products chases the bourgeoisie over the whole surface of the globe. It must nestle everywhere, settle everywhere, establish [connections] everywhere.

The bourgeoisie has through its exploitation of the world-market given a cosmopolitan character to production and consumption in every country. To the great chagrin of Reactionists, it has drawn from under the feet of industry the national ground on which it stood. All old-established national industries have been destroyed or are daily being destroyed. They are dislodged by new industries, whose introduction becomes a life and death question for all civilised nations, by industries that no longer work up indigenous raw material, but rare material drawn from the remotest zones; industries whose products are consumed, not only at home, but in every quarter of the globe. In place of the old wants, satisfied by the productions of the country, we find new wants, requiring for their satisfaction the products of distant lands and climes. In place of the old local and national seclusion and self-sufficiency, we have intercourse in every direction, universal interdependence of nations. And as in material, so also in intellectual production. The intellectual creations of individual nations become common property. National one-sidedness and narrow-mindedness become more and more impossible, and from the numerous national and local literatures there arises a world literature.

The bourgeoisie, by the rapid improvement of all instruments of production, by the immensely facilitated means of communication, draws all, even the most barbarian, nations into civilization. The cheap prices of its commodities are the heavy artillery with which it batters down all Chinese walls, with which it forces the barbarians' intensely obstinate hatred of foreigners to capitulate. It compels all nations, on pain of extinction, to adopt the bourgeois mode of production; it compels them to introduce what it calls civilization into their midst, i.e., to become bourgeois themselves. In a word, it creates a world after its own image.

The bourgeoisie has subjected the country to the rule of the towns. It has created enormous cities, has greatly increased the urban population as compared with the rural, and has thus rescued a considerable part of the population from the idiocy of rural life. Just as it has made the country dependent on the towns, so it has made barbarian and semi-barbarian countries dependent on the civilized ones, nations of peasants on nations of bourgeois, the East on the West.

The bourgeoisie keeps more and more doing away with the scattered state of the population, of the means of production, and of property. It has agglomerated population, centralized means of production, and has concentrated property in a few hands. The necessary consequence of this was political centralization. Independent, or but loosely connected provinces, with separate interests, laws, governments and systems of taxation, became lumped together in one nation, with one government, one code of laws, one national class-interest, one frontier and one customs-tariff.

The bourgeoisie, during its rule of scarce one hundred years, has created more massive and more colossal productive forces than have all preceding generations together. Subjection of Nature's forces to man, machinery, application of chemistry to industry and agriculture, steam-navigation, railways, electric telegraphs, clearing of whole continents for cultivation, canalization of rivers, whole populations conjured out of the ground—what earlier century had even a presentiment that such productive forces slumbered in the lap of social labor?

We see then: The means of production and of exchange on whose foundation the bourgeoisie built itself up, were generated in feudal society. At a certain stage in the development of these means of production and of exchange, the conditions under which feudal society produced and exchanged, the feudal organization of agriculture and manufacturing industry, in one word, the feudal relations of property became no longer compatible with the already developed productive forces; they became so many fetters. They had to burst asunder; they were burst asunder.

Into their places stepped free competition, accompanied by a social and political constitution adapted to it, and by the economical and political sway of the bourgeois class.

A similar movement is going on before our own eyes. Modern bourgeois society with its relations of production, of exchange and of property, a society that has conjured up such gigantic means of production and of exchange, is like the sorcerer, who is no longer able to control the powers of the nether world whom he has called up by his spells. For many a decade past the history of industry and commerce is but the history of the revolt of modern productive forces against modern conditions of production, against the property relations that are the conditions for the existence of the bourgeoisie and of its rule. It is enough to mention the commercial crises that by their periodical return put on its trial, each time more threateningly, the existence of the entire bourgeois society. In these crises a great part not only of the existing products, but also of the previously created productive forces, are periodically destroyed. In these crises there breaks out an epidemic that, in all earlier epochs, would have seemed an absurdity—the epidemic of overproduction. Society suddenly finds itself put back into a state of momentary barbarism; it appears as if a famine, a universal war of devastation had cut off the supply of every means of subsistence; industry and commerce seem to be destroyed; and why? Because there is too much civilization, too much means of subsistence, too much industry, too much commerce. The productive forces at the disposal of society no longer tend to further the development of the conditions of bourgeois property; on the contrary, they have become too powerful for these conditions, by which they are fettered, and so

soon as they overcome these fetters, they bring disorder into the whole of bourgeois society, endanger the existence of bourgeois property. The conditions of bourgeois society are too narrow to comprise the wealth created by them. And how does the bourgeoisie get over these crises? On the one hand by enforced destruction of a mass of productive forces; on the other, by the conquest of new markets, and by the more thorough exploitation of the old ones. That is to say, by paving the way for more extensive and more destructive crises, and by diminishing the means whereby crises are prevented.

The weapons with which the bourgeoisie felled feudalism to the ground are now turned against the bourgeoisie itself. But not only has the bourgeoisie forged the weapons that bring death to itself; it has also called into existence the men who are to wield those weapons—the modern working class—the proletarians.

In proportion as the bourgeoisie, i.e., capital, is developed, in the same proportion is the proletariat, the modern working class, developed, a class of laborers, who live only so long as they find work, and who find work only so long as their labor increases capital. These laborers, who must sell themselves piecemeal, are a commodity, like every other article of commerce, and are consequently exposed to all the vicissitudes of competition, to all the fluctuations of the market.

Owing to the extensive use of machinery and to division of labor, the work of the proletarians has lost all individual character, and, consequently, all charm for the workman. He becomes an appendage of the machine, and it is only the most simple, most monotonous, and most easily acquired knack that is required of him. Hence, the cost of production of a workman is restricted, almost entirely, to the means of subsistence that he requires for his maintenance, and for the propagation of his race. But the price of a commodity, and also of labor, is equal to its cost of production. In proportion, therefore, as the repulsiveness of the work increases, the wage decreases. Nay more, in proportion as the use of machinery and division of labor increases, in the same proportion the burden of toil also increases, whether by prolongation of the working hours, by increase of the work enacted in a given time, or by increased speed of the machinery, etc.

Modern industry has converted the little workshop of the patriarchal master in the great factory of the industrial capitalist. Masses of laborers, crowded into the factory, are organized like soldiers. As privates of the industrial army they are placed under the command of a perfect hierarchy of officers and sergeants. Not only are they the slaves of the bourgeois class, and of the bourgeois State, they are daily and hourly enslaved by the machine, by the overlooker, and, above all, by the individual bourgeois manufacturer himself. The more openly this despotism proclaims gain to be its end and aim, the more petty, the more hateful and the more embittering it is.

The less the skill and exertion or strength implied in manual labor, in other words, the more modern industry becomes developed, the more is the labor of men superseded by that of women. Differences of age and sex have no longer any distinctive social validity for the working class. All are instruments of labor, more or less expensive to use, according to their age and sex.

No sooner is the exploitation of the laborer by the manufacturer, so far, at an end, that he receives his wages in cash, than he is set upon by the other portions of the bourgeoisie, the landlord, the shopkeeper, the pawnbroker, etc.

The lower strata of the middle class—the small tradespeople, shopkeepers, and retired tradesmen generally, the handicraftsmen and peasants—all these sink gradually into the proletariat, partly because their diminutive capital does not suffice for the scale on which Modern Industry is carried on, and is swamped in the competition with the large capitalists, partly because their specialised skill is rendered worthless by new methods of production. Thus the proletariat is recruited from all classes of the population.

The proletariat goes through various stages of development. With its birth begins its struggle with the bourgeoisie. At first the contest is carried on by individual laborers, then by the workpeople of a factory, then by the operatives of one trade, in one locality, against the individual bourgeois who directly exploits them. They direct their attacks not against the bourgeois conditions of production, but against the instruments of production themselves; they destroy imported wares that compete with their labor, they smash to pieces machinery, they set factories ablaze, they seek to restore by force the vanished status of the workman of the Middle Ages.

At this stage the laborers still form an incoherent mass scattered over the whole country, and broken up by their mutual competition. If any where they unite to form more compact bodies, this is not yet the consequence of their own active union, but of the union of the bourgeoisie, which class, in order to attain its own political ends, is compelled to set the whole proletariat in motion, and is moreover yet, for a time, able to do so. At this stage, therefore, the proletarians do not fight their enemies, but the enemies of their enemies, the remnants of absolute monarchy, the landowners, the non-industrial bourgeois, the petty bourgeoisie. Thus the whole historical movement is concentrated in the hands of the bourgeoisie; every victory so obtained is a victory for the bourgeoisie.

But with the development of industry the proletariat not only increases in number, it becomes concentrated in greater masses, its strength grows, and it feels that strength more. The various interests and conditions of life within the ranks of the proletariat are more and more equalized, in proportion as machinery obliterates all distinctions of labor, and nearly everywhere reduces wages to the same low level. The growing competition among the bourgeois, and the resulting commercial crises, make the wages of the workers ever more fluctuating. The unceasing improvement of machinery, ever more rapidly developing, makes their livelihood more and more precarious, the collisions between individual workmen and individual bourgeois take more and more the character of collisions between two classes. Thereupon the workers begin to form combinations (Trades' Unions) against the bourgeois; they club together in order to keep up the rate of wages; they found permanent associations in order to make provision beforehand for these occasional revolts. Here and there the contest breaks out into riots.

Now and then the workers are victorious, but only for a time. The real fruit of their battles lies, not in the immediate result, but in the ever expanding union of the workers. This union is helped on by the improved means of communication that are created by modern industry, and that place the workers of different localities in contact with one another. It was just this contact that was needed to centralise the numerous local struggles, all of the same character, into one national struggle between classes. But every class struggle is a political struggle. And that union, to attain which the burghers of the Middle Ages, with their miserable highways, required centuries, the modern proletarians, thanks to railways, achieve in a few years.

This organization of the proletarians into a class, and consequently into a political party, is continually being upset again by the competition between the workers themselves. But it ever rises up again, stronger, firmer, mightier. It compels legislative recognition of particular interests of the workers, by taking advantage of the divisions among the bourgeoisie itself. Thus the ten–hours' bill in England was carried.

Altogether collisions between the classes of the old society further, in many ways, the course of development of the proletariat. The bourgeoisie finds itself involved in a constant battle. At first with the aristocracy; later on, with those portions of the bourgeoisie itself, whose interests have become antagonistic to the progress of industry; at all times, with the bourgeoisie of foreign countries.

In all these battles it sees itself compelled to appeal to the proletariat, to ask for its help, and thus, to drag it into the political arena. The bourgeoisie itself, therefore, supplies the proletariat with its own elements of political and general education, in other words, it furnishes the proletariat with weapons for fighting the bourgeoisie.

Further, as we have already seen, entire sections of the ruling classes are, by the advance of industry, precipitated into the proletariat, or are at least threatened in their conditions of existence. These also supply the proletariat with fresh elements of enlightenment and progress.

Finally, in times when the class-struggle nears the decisive hour, the process of dissolution going on within the ruling class, in fact within the whole range of old society, assumes such a violent, glaring character, that a small section of the ruling class cuts itself adrift, and joins the revolutionary class, the class that holds the future in its hands. Just as, therefore, at an earlier period, a section of the nobility went over to the bourgeoisie, so now a portion of the bourgeoisie goes over to the proletariat, and in particular, a portion of the bourgeois ideologists, who have raised themselves to the level of comprehending theoretically the historical movements as a whole.

Of all the classes that stand face to face with the bourgeoisie today, the proletariat alone is a really revolutionary class. The other classes decay and finally disappear in the face of modern industry; the proletariat is its special and essential product.

The lower-middle class, the small manufacturer, the shopkeeper, the artisan, the peasant, all these fight against the bourgeoisie, to save from extinction their existence as fractions of the middle class. They are therefore not revolutionary, but conservative. Nay more, they are reactionary, for they try to roll back the wheel of history. If by chance they are revolutionary, they are so only in view of their impending transfer into the proletariat, they thus defend not their present, but their future interests, they desert their own standpoint to place themselves at that of the proletariat.

The "dangerous class," the social scum, that passively rotting mass thrown off by the lowest layers of old society, may, here and there, be swept into the movement by a proletarian revolution; its conditions of life, however, prepare it far more for the part of a bribed tool of reactionary intrigue.

In the conditions of the proletariat, those of old society at large are already virtually swamped. The proletarian is without property; his relation to his wife and children has no longer anything in common with the bourgeois family-relations; modern industrial labor, modern subjection to capital, the same in England as in France, in America as in Germany, has stripped him of every trace of national character. Law, morality, religion, are to him so many bourgeois prejudices, behind which lurk in ambush just as many bourgeois interests.

All the preceding classes that got the upper hand sought to fortify their already acquired status by subjecting society at large to their conditions of appropriation. The proletarians cannot become masters of the productive forces of society, except by abolishing their own previous mode of appropriation, and thereby also every other previous mode of appropriation. They have nothing of their own to secure and to fortify; their mission is to destroy all previous securities for, and insurances of, individual property.

All previous historical movements were movements of minorities, or in the interest of minorities. The proletarian movement is the self-conscious, independent movement of the immense majority, in the interest of the immense majority. The proletariat, the lowest stratum of our present society, cannot stir, cannot raise itself up, without the whole superincumbent strata of official society being sprung into the air.

Though not in substance, yet in form, the struggle of the proletariat with the bourgeoisie is at first a national struggle. The proletariat of each country must, of course, first of all settle matters with its own bourgeoisie.

In depicting the most general phases of the development of the proletariat, we traced the more or less veiled civil war, raging within existing society, up to the point where that war breaks out into open revolution, and where the violent overthrow of the bourgeoisie lays the foundation for the sway of the proletariat.

Hitherto, every form of society has been based, as we have already seen, on the antagonism of oppressing and oppressed classes. But in order to oppress a class, certain conditions must be assured to it under which it can, at least, continue its slavish existence. The serf, in the period of serfdom, raised himself to membership in the commune, just as the petty bourgeois, under the yoke of feudal absolutism, managed to develop into a bourgeois. The modern laborer, on the contrary, instead of rising with the progress of industry, sinks deeper and deeper below the conditions of existence of his own class. He becomes a pauper, and pauperism develops more rapidly than population and wealth. And here it becomes evident, that the bourgeoisie is unfit any longer to be the ruling class in society, and to impose its conditions of existence upon society as an overriding law. It is unfit to rule, because it is incompetant to assure an existence to its slave within his slavery, because it cannot help letting him sink into such a state, that it has to feed him, instead of being fed by him. Society can no longer live under this bourgeoisie, in other words, its existence is no longer compatible with society.

The essential condition for the existence, and for the sway of the bourgeois class, is the formation and augmentation of capital; the condition for capital is wage-labor. Wage-labor rests exclusively on competition between the laborers. The advance of industry, whose involuntary promoter is the bourgeoisie, replaces the isolation of the laborers, due to competition, by their involuntary combination, due to association. The development of Modern Industry, therefore, cuts from under its feet the very foundation on which the bourgeoisie produces and appropriates products. What the bourgeoisie therefore produces, above all, are its own gravediggers. Its fall and the victory of the proletariat are equally inevitable.

NOTES

1. By *bourgeoisie* is meant the class of modern capitalists, owners of the means of social production and employers of wage-labor. By *proletariat*, the class of modern wage-laborers who, having no means of production of their own, are reduced to selling their labor-power in order to live.

2. That is, all written history. In 1847, the prehistory of society, the social organization existing previous to recorded history, was all but unknown. Since then, Haxthausen discovered common ownership of land in Russia. Maurer proved it to be the social foundation from which all Teutonic races started in

history, and by and by village communities were found to be, or to have been the primitive form of society everywhere from India to Ireland. The inner organization of this primitive Communistic society was laid bare, in its typical form, by Morgan's crowning discovery of the true nature of the gens and its relationship to the tribe. With the dissolution of these primaeval communities society begins to be differentiated into separate and finally antagonistic classes. I have attempted to retrace this process of dissolution in "Der Ursprung der Familie des, Privatelgenthums und des Staats," 2d ed., Stuttgart 1886.

3. Guild-master that is a full member of a guild, a master within, not head of, a guild.

4. "Commune" was the name taken, in France, by the nascent towns even before they had conquered from their feudal lords and masters, local selfgovernment and political rights as "the Third Estate." Generally speaking, for the economical development of the bourgeoisie, England is here taken as the typical country, for its political development, France.

DISCUSSION QUESTIONS

1. Karl Marx and Friedrich Engels see class conflict as inherent in capitalist societies. What changes in class relations do they envision over time? Do you see evidence of this in the contemporary world?

2. What do Marx and Engels mean when they say that the bourgeoisie class turns all social relations into money relations? What evidence do you see of this?

3. Marx and Engels were writing more than 150 years ago. What evidence do you see now of increasing consumerism and how this shapes the values of people in the United States? What effect does it have on the class system?

4. According to Marx and Engels, the bourgeoisie is driven to create ever-new markets, and it draws nations of the world into this capitalist market. If Marx and Engels were writing in the twenty-first century, what kinds of similar phenomena might they include in a newly written essay?

INTERNET RESOURCES

Suggested Web URLs for Further Study

http://www.marxists.org
This website contains an encyclopedia of Marxism, reference writers, a history archive, and links to more Marxist writers.

Virtual Society

For further information on this subject, including links to relevant websites, go to the Wadsworth Sociology homepage at *http://www.cengage.com/sociology*.

4

Body Ritual Among the Nacirema

HORACE MINER

The anthropologist has become so familiar with the diversity of ways in which different peoples behave in similar situations that he is not apt to be surprised by even the most exotic customs. In fact, if all of the logically possible combinations of behavior have not been found somewhere in the world, he is apt to suspect that they must be present in some yet undescribed tribe. This point has, in fact, been expressed with respect to clan organization by Murdock (1949: 71). In this light, the magical beliefs and practices of the Nacirema present such unusual aspects that it seems desirable to describe them as an example of the extremes to which human behavior can go.

Professor Linton first brought the ritual of the Nacirema to the attention of anthropologists twenty years ago (1936: 326), but the culture of this people is still very poorly understood. They are a North American group living in the territory between the Canadian Cree, the Yaqui and Tarahumare of Mexico, and the Carib and Arawak of the Antilles. Little is known of their origin, although tradition states that they came from the east. According to Nacirema mythology, their nation was originated by a culture hero, Notgnihsaw, who is otherwise known for two great feats of strength—the throwing of a piece of wampum across the river Pa-To-Mac and the chopping down of a cherry tree in which the Spirit of Truth resided.

Nacirema culture is characterized by a highly developed market economy which has evolved in a rich natural habitat. While much of the people's time is devoted to economic pursuits, a large part of the fruits of these labors and a considerable portion of the day are spent in ritual activity. The focus of this activity is the human body, the appearance and health of which loom as a dominant

SOURCE: "Body Ritual Among the Nacirema," by Horace Miner, reproduced by permission of the American Anthropological Association from *American Anthropologist* 58:3, June 1956. Not for further reproduction.

concern in the ethos of the people. While such concern is certainly not unusual, its ceremonial aspects and associated philosophy are unique.

The fundamental belief underlying the whole system appears to be that the human body is ugly and that its natural tendency is to debility and disease. Incarcerated in such a body, man's only hope is to avert these characteristics through the use of the powerful influences of ritual and ceremony. Every household has one or more shrines devoted to this purpose. The more powerful individuals in this society have several shrines in their houses, and, in fact, the opulence of a house is often referred to in terms of the number of such ritual centers it possesses. Most houses are of wattle and daub construction, but the shrine rooms of the more wealthy are walled with stone. Poorer families imitate the rich by applying pottery plaques to their shrine walls.

While each family has at least one such shrine, the rituals associated with it are not family ceremonies but are private and secret. The rites are normally only discussed with children, and then only during the period when they are being initiated into these mysteries. I was able, however, to establish sufficient rapport with the natives to examine these shrines and to have the rituals described to me.

The focal point of the shrine is a box or chest which is built into the wall. In this chest are kept the many charms and magical potions without which no native believes he could live. These preparations are secured from a variety of specialized practitioners. The most powerful of these are the medicine men, whose assistance must be rewarded with substantial gifts. However, the medicine men do not provide the curative potions for their clients, but decide what the ingredients should be and then write them down in an ancient and secret language. This writing is understood only by the medicine men and by the herbalists who, for another gift, provide the required charm.

The charm is not disposed of after it has served its purpose, but is placed in the charm-box of the household shrine. As these magical materials are specific for certain ills, and the real or imagined maladies of the people are many, the charm-box is usually full to overflowing. The magical packets are so numerous that people forget what their purposes were and fear to use them again. While the natives are very vague on this point, we can only assume that the idea in retaining all the old magical materials is that their presence in the charm-box, before which the body rituals are conducted, will in some way protect the worshipper.

Beneath the charm-box is a small font. Each day every member of the family, in succession, enters the shrine room, bows his head before the charm-box, mingles different sorts of holy water in the font, and proceeds with a brief rite of ablution. The holy waters are secured from the Water Temple of the community, where the priests conduct elaborate ceremonies to make the liquid ritually pure.

In the hierarchy of magical practitioners, and below the medicine men in prestige, are specialists whose designation is best translated "holy-mouthmen." The Nacirema have an almost pathological horror of and fascination with the mouth, the condition of which is believed to have a supernatural influence on all social relationships. Were it not for the rituals of the mouth, they believe that their teeth would fall out, their gums bleed, their jaws shrink, their friends desert them, and their lovers reject them. They also believe that a strong relationship exists between oral and moral characteristics. For example, there is a ritual ablution of the mouth for children which is supposed to improve their moral fiber.

The daily body ritual performed by everyone includes a mouth-rite. Despite the fact that these people are so punctilious about care of the mouth, this rite involves a practice which strikes the uninitiated stranger as revolting. It was reported to me that the ritual consists of inserting a small bundle of hog hairs into the mouth, along with certain magical powders, and then moving the bundle in a highly formalized series of gestures.

In addition to the private mouth-rite, the people seek out a holy-mouth-man once or twice a year. These practitioners have an impressive set of paraphernalia, consisting of a variety of augers, awls, probes, and prods. The use of these objects in the exorcism of the evils of the mouth involves almost unbelievable ritual torture of the client. The holy-mouth-man opens the client's mouth and, using the above-mentioned tools, enlarges any holes which decay may have created in the teeth. Magical materials are put into these holes. If there are no naturally occurring holes in the teeth, large sections of one or more teeth are gouged out so that the supernatural substance can be applied. In the client's view, the purpose of these ministrations is to arrest decay and to draw friends. The extremely sacred and traditional character of the rite is evident in the fact that the natives return to the holy-mouth-man year after year, despite the fact that their teeth continue to decay.

It is to be hoped that, when a thorough study of the Nacirema is made, there will be careful inquiry into the personality structure of these people. One has but to watch the gleam in the eye of a holy-mouth-man, as he jabs an awl into an exposed nerve, to suspect that a certain amount of sadism is involved. If this can be established, a very interesting pattern emerges, for most of the population shows definite masochistic tendencies. It was to these that Professor Linton referred in discussing a distinctive part of the daily body ritual which is performed only by men. This part of the rite involves scraping and lacerating the surface of the face with a sharp instrument. Special women's rites are performed only four times during each lunar month, but what they lack in frequency is made up in barbarity. As part of this ceremony, women bake their heads in small ovens for about an hour. The theoretically interesting point is that what seems to be a preponderantly masochistic people have developed sadistic specialists.

The medicine men have an imposing temple, or *latipso*, in every community of any size. The more elaborate ceremonies required to treat very sick patients can only be performed at this temple. These ceremonies involve not only the thaumaturge but a permanent group of vestal maidens who move sedately about the temple chambers in distinctive costume and headdress.

The *latipso* ceremonies are so harsh that it is phenomenal that a fair proportion of the really sick natives who enter the temple ever recover. Small children whose indoctrination is still incomplete have been known to resist attempts to take them to the temple because "that is where you go to die." Despite this fact, sick adults are not only willing but eager to undergo the protracted ritual purification, if they can afford to do so. No matter how ill the supplicant or how grave the emergency, the guardians of many temples will not admit a client if he cannot give a rich gift to the custodian. Even after one has gained admission and survived the ceremonies, the guardians will not permit the neophyte to leave until he makes still another gift.

The supplicant entering the temple is first stripped of all his or her clothes. In everyday life the Nacirema avoids exposure of his body and its natural functions. Bathing and excretory acts are performed only in the secrecy of the household shrine, where they are ritualized as part of the body-rites. Psychological shock results from the fact that body secrecy is suddenly lost upon entry into the *latipso*. A man, whose own wife has never seen him in an excretory act, suddenly finds himself naked and assisted by a vestal maiden while he performs his natural functions into a sacred vessel. This sort of ceremonial treatment is necessitated by the fact that the excreta are used by a diviner to ascertain the course and nature of the client's sickness. Female clients, on the other hand, find their naked bodies are subjected to the scrutiny, manipulation, and prodding of the medicine men.

Few supplicants in the temple are well enough to do anything but lie on their hard beds. The daily ceremonies, like the rites of the holy-mouth-men, involve discomfort and torture. With ritual precision, the vestals awaken their miserable charges each dawn and roll them about on their beds of pain while performing ablutions, in the formal movements of which the maidens are highly trained. At other times they insert magic wands in the supplicant's mouth or force him to eat substances which are supposed to be healing. From time to time the medicine men come to their clients and jab magically treated needles into their flesh. The fact that these temple ceremonies may not cure, and may kill, the neophyte, in no way decreases the people's faith in the medicine men.

There remains one other kind of practitioner, known as a "listener." This witch-doctor has the power to exorcise the devils that lodge in the heads of people who have been bewitched. The Nacirema believe that parents bewitch their own children. Mothers are particularly suspected of putting a curse on children while teaching them the secret body rituals. The counter-magic of the witch-doctor is unusual in its lack of ritual. The patient simply tells the "listener" all his troubles and fears, beginning with the earliest difficulties he can remember. The memory displayed by the Nacirema in these exorcism sessions is truly remarkable. It is not uncommon for the patient to bemoan the rejection he felt upon being weaned as a babe, and a few individuals even see their troubles going back to the traumatic effects of their own birth.

In conclusion, mention must be made of certain practices which have their base in native esthetics but which depend upon the pervasive aversion to the natural body and its functions.

There are ritual fasts to make fat people thin and ceremonial feasts to make thin people fat. Still other rites are used to make women's breasts larger if they are small, and smaller if they are large. General dissatisfaction with breast shape is symbolized in the fact that the ideal form is virtually outside the range of human variation. A few women afflicted with almost inhuman hyper-mammary development are so idolized that they make a handsome living by simply going from village to village and permitting the natives to stare at them for a fee.

Reference has already been made to the fact that excretory functions are ritualized, routinized, and relegated to secrecy. Natural reproductive functions are similarly distorted. Intercourse is taboo as a topic and scheduled as an act. Efforts are made to avoid pregnancy by the use of magical materials or by limiting intercourse to certain phases of the moon. Conception is actually very infrequent. When pregnant, women dress so as to hide their condition. Parturition takes place in secret, without friends or relatives to assist, and the majority of women do not nurse their infants.

Our review of the ritual life of the Nacirema has certainly shown them to be a magic-ridden people. It is hard to understand how they have managed to exist so long under the burdens which they have imposed upon themselves. But even such exotic customs as these take on real meaning when they are viewed with the insight provided by Malinowski when he wrote (1948: 70):

> Looking from far and above, from our high places of safety in the developed civilization, it is easy to see all the crudity and irrelevance of magic. But without its power and guidance early man could not have mastered his practical difficulties as he has done, nor could man have advanced to the higher stages of civilization.

REFERENCES

Linton, R. 1936. *The study of man*. New York: Appleton-Century.

Malinowski, B. 1948. *Magic, science and religion*. Glencoe, IL: Free Press.

Murdock, G. P. 1949. *Social structure*. New York: Macmillan.

DISCUSSION QUESTIONS

1. Horace Miner's satirical piece is a good example of what Peter Berger calls "debunking." Miner debunks the ordinary, everyday cultural practices of Americans. Identify a different cultural practice and then write a similar essay debunking it as if you were looking at it as a complete outsider. What do you learn from doing so?

2. What other kinds of rituals are common in American society—other than the cleansing rituals about which Miner writes? What function do such rituals have in American culture—other than their alleged purpose?

3. Miner presents the practices of the Nacirema as if they were magic. How might you see the same practices if you looked at them as if they stemmed solely from a scientific and technological worldview? What if you looked at them as a form of sport or drama? How does your view of behavior influence what you see?

INTERNET RESOURCES

Suggested Web URLs for Further Study

http://www.beadsland.com/nacirema/#anthro
This website covers further related readings on the Nacirema.

Virtual Society

For further information on this subject, including links to relevant websites, go to the Wadsworth Sociology homepage at *http://www.cengage.com/sociology*.

5

Lynch Law in America

IDA B. WELLS-BARNETT

Our country's national crime is lynching. It is not the creature of an hour, the sudden outburst of uncontrolled fury, or the unspeakable brutality of an insane mob. It represents the cool, calculating deliberation of intelligent people who openly avow that there is an "unwritten law" that justifies them in putting human beings to death without complaint under oath, without trial by jury, without opportunity to make defense, and without right of appeal. The "unwritten law" first found excuse with the rough, rugged, and determined man who left the civilized centers of eastern States to seek for quick returns in the gold-fields of the far West. Following in uncertain pursuit of continually eluding fortune … the hardships of mountain travel, and the constant terror of border State outlaws. Naturally, they felt slight toleration for traitors in their own ranks. It was enough to fight the enemies from without; woe to the foe within! Far removed from and entirely without protection of the courts of civilized life, these fortune-seekers made laws to meet their varying emergencies. The thief who stole a horse, the bully who "jumped" a claim, was a common enemy. If caught he was promptly tried, and if found guilty was hanged to the tree under which the court convened.

Those were busy days of busy men. They had no time to give the prisoner a bill of exception or stay of execution. The only way a man had to secure a stay of execution was to behave himself. Judge Lynch was original in methods but exceedingly effective in procedure. He made the charge, impaneled the jurors, and directed the execution. When the court adjourned, the prisoner was dead. Thus lynch law held sway in the far West until civilization spread into the Territories and the orderly processes of law took its place. The emergency no longer existing, lynching gradually disappeared from the West.

But the spirit of mob procedure seemed to have fastened itself upon the lawless classes, and the grim process that at first was invoked to declare justice was made the excuse to wreak vengeance and cover crime. It next appeared in the South, where centuries of Anglo-Saxon civilization had made effective all the safeguards of court procedure. No emergency called for lynch law. It asserted its sway in defiance of law and in favor of anarchy. There it has flourished ever since, marking the thirty years of its existence with the inhuman butchery of more than ten thousand men, women, and children by shooting, drowning, hanging, and burning them alive. Not only this, but so potent is the force of example that the lynching mania has spread throughout the North and middle West. It is now no uncommon thing to read of lynchings north of Mason and Dixon's line, and those most responsible for this fashion gleefully point to these instances and assert that the North is no better than the South.

This is the work of the "unwritten law" about which so much is said, and in whose behest butchery is made a pastime and national savagery condoned. The first statute of this "unwritten law" was written in the blood of thousands of brave men who thought that a government that was good enough to create a citizenship was strong enough to protect it. Under the authority of a national law that gave every citizen the right to vote, the newly-made citizens chose to exercise their suffrage. But the reign of the national law was short-lived and illusionary. Hardly had the sentences dried upon the statute-books before one Southern State after another raised the cry against "[N]egro* domination" and proclaimed there was an "unwritten law" that justified any means to resist it.

The method then inaugurated was the outrages by the "red-shirt" bands of Louisiana, South Carolina, and other Southern States, which were succeeded by the Ku-Klux Klans. These advocates of the "unwritten law" boldly avowed their purpose to intimidate, suppress, and nullify the [N]egro's right to vote. In support of its plans the Ku-Klux Klans, the "red-shirt" and similar organizations proceeded to beat, exile, and kill [N]egroes until the purpose of their organization was accomplished and the

supremacy of the "unwritten law" was effected. Thus lynchings began in the South, rapidly spreading into the various States until the national law was nullified and the reign of the "unwritten law" was supreme. Men were taken from their homes by "red-shirt" bands and stripped, beaten, and exiled; others were assassinated when their political prominence made them obnoxious to their political opponents; while the Ku-Klux barbarism of election days, reveling in the butchery of thousands of colored voters, furnished records in Congressional investigations that are a disgrace to civilization.

The alleged menace of universal suffrage having been avoided by the absolute suppression of the [N]egro vote, the spirit of mob murder should have been satisfied and the butchery of [N]egroes should have ceased. But men, women, and children were the victims of murder by individuals and murder by mobs, just as they had been when killed at the demands of the "unwritten law" to prevent "[N]egro domination." Negroes were killed for disputing over terms of contracts with their employers. If a few barns were burned some colored man was killed to stop it. If a colored man resented the imposition of a white man and the two came to blows, the colored man had to die, either at the hands of the white man then and there or later at the hands of a mob that speedily gathered. If he showed a spirit of courageous manhood he was hanged for his pains, and the killing was justified by the declaration that he was a "saucy nigger." Colored women have been murdered because they refused to tell the mobs where relatives could be found for "lynching bees." Boys of fourteen years have been lynched by white representatives of American civilization. In fact, for all kinds of offenses—and, for no offenses—from murders to misdemeanors, men and women are put to death without judge or jury;

---

*When Ida B. Wells-Barnett wrote "Lynch Law in America," it had not yet become standard usage in America (and later the world) for the word Negro to be capitalized when written. The editor of this edition of Mrs. Wells-Barnett's piece decided to alter her spelling of the word since she was using it in 1900, well before the NAACP, founded in 1909, had campaigned for the capitalization of the word "Negro" and the *New York Times* announced in an editorial in 1930 that the word Negro would be among those words it would capitalize. From then on the word has been spelled with a capital "N." Current history tells us the word Negro was used interchangeably with the word "Colored" and "Afro-American"; around the middle of the last century, both words were supplanted by "Black," "African American," and "African."

so that, although the political excuse was no longer necessary, the wholesale murder of human beings went on
just the same. A new name was given to the killings
and a new excuse was invented for so doing.

Again the aid of the "unwritten law" is invoked, and again it comes to the rescue. During
the last ten years a new statute has been added to
the "unwritten law." This statute proclaims that
for certain crimes or alleged crimes no [N]egro
shall be allowed a trial; that no white woman shall
be compelled to charge an assault under oath or to
submit any such charge to the investigation of a
court of law. The result is that many men have
been put to death whose innocence was afterward
established; and to-day [sic], under this reign of the
"unwritten law," no colored man, no matter what
his reputation, is safe from lynching if a white
woman, no matter what her standing or motive,
cares to charge him with insult or assault.

It is considered a sufficient excuse and reasonable justification to put a prisoner to death under
this "unwritten law" for the frequently repeated
charge that these lynching horrors are necessary to
prevent crimes against women. The sentiment of
the country has been appealed to, in describing
the isolated condition of white families in thickly
populated [N]egro districts; and the charge is made
that these homes are in as great danger as if they
were surrounded by wild beasts. And the world
has accepted this theory without let or hindrance.
In many cases there has been open expression that
the fate meted out to the victim was only what he
deserved. In many other instances there has been a
silence that says more forcibly than words can pro-
claim it that it is right and proper that a human
being should be seized by a mob and burned to
death upon the unsworn and the uncorroborated
charge of his accuser. No matter that our laws presume every man innocent until he is proved guilty;
no matter that it leaves a certain class of individuals
completely at the mercy of another class; no matter
that it encourages those criminally disposed to
blacken their faces and commit any crime in the
calendar so long as they can throw suspicion on
some [N]egro, as is frequently done, and then lead
a mob to take his life; no matter that mobs make a

farce of the law and a mockery of justice; no matter
that hundreds of boys are being hardened in crime
and schooled in vice by the repetition of such scenes
before their eyes—if a white woman declares herself
insulted or assaulted, some life must pay the penalty,
with all the horrors of the Spanish Inquisition and all
the barbarism of the Middle Ages. The world looks
on and says it is well.

Not only are two hundred men and women
put to death annually, on the average, in this country by mobs, but these lives are taken with the
greatest publicity. In many instances the leading citizens aid and abet by their presence when they do
not participate, and the leading journals inflame the
public mind to the lynching point with scare-head
articles and offers of rewards.

Whenever a burning is advertised to take place,
the railroads run excursions, photographs are taken,
and the same jubilee is indulged in that characterized
the public hangings of one hundred years ago. There
is, however, this difference: in those old days the
multitude that stood by was permitted only to guy
or jeer. The nineteenth century lynching mob cuts
off ears, toes, and fingers, strips off flesh, and distri-
butes portions of the body as souvenirs among the
crowd. If the leaders of the mob are so minded, coal-
oil is poured over the body and the victim is then
roasted to death. This has been done in Texarkana
and Paris, Tex., in Bardswell, Ky., and in Newman,
Ga. In Paris the officers of the law delivered the prisoner to the mob. The mayor gave the school children a holiday and the railroads ran excursion
trains so that the people might see a human being
burned to death. In Texarkana, the year before,
men and boys amused themselves by cutting off
strips of flesh and thrusting knives into their helpless victim. At Newman, Ga., of the present year,
the mob tried every conceivable torture to compel
the victim to cry out and confess, before they set
fire to the faggots that burned him. But their trouble was all in vain—he never uttered a cry, and
they could not make him confess.

This condition of affairs were brutal enough
and horrible enough if it were true that lynchings
occurred only because of the commission of crimes
against women—as is constantly declared by

ministers, editors, lawyers, teachers, statesmen, and even by women themselves. It has been to the interest of those who did the lynching to blacken the good name of the helpless and defenseless victims of their hate. For this reason they publish at every possible opportunity this excuse for lynching, hoping thereby not only to palliate their own crime but at the same time to prove the [N]egro a moral monster and unworthy of the respect and sympathy of the civilized world. But this alleged reason adds to the deliberate injustice of the mob's work. Instead of lynchings being caused by assaults upon women, the statistics show that not one-third of the victims of lynchings are even charged with such crimes. The *Chicago Tribune*, which publishes annually lynching statistics, is authority for the following:

In 1892, when lynching reached high-water mark, there were 241 persons lynched. The entire number is divided among the following States:

Alabama	22	Montana	4
Arkansas	25	New York	1
California	3	North Carolina	5
Florida	11	North Dakota	1
Georgia	17	Ohio	3
Idaho	8	South Carolina	5
Illinois	1	Tennessee	28
Kansas	3	Texas	15
Kentucky	9	Virginia	7
Louisiana	29	West Virginia	5
Maryland	1	Wyoming	9
Mississippi	16	Arizona Territory	3
Missouri	6	Oklahoma	2

Of this number, 160 were of [N]egro descent. Four of them were lynched in New York, Ohio, and Kansas; the remainder were murdered in the South. Five of this number were females. The charges for which they were lynched cover a wide range. They are as follows:

Rape	46	Attempted rape	11
Murder	58	Suspected robbery	4
Rioting	3	Larceny	1
Race Prejudice	6	Self-defense	1
No cause given	4	Insulting women	2
Incendiarism	6	Desperadoes	6
Robbery	6	Fraud	1
Assault and battery	1	Attempted murder	2
No offense stated, boy and girl			2

In the case of the boy and girl above referred to, their father, named Hastings, was accused of the murder of a white man. His fourteen-year-old daughter and sixteen-year-old son were hanged and their bodies filled with bullets; then the father was also lynched. This occurred in November, 1892, at Jonesville, La.

Indeed, the record for the last twenty years shows exactly the same or a smaller proportion who have been charged with this horrible crime. Quite a number of the one-third alleged cases of assault that have been personally investigated by the writer have shown that there was no foundation in fact for the charges; yet the claim is not made that there were no real culprits among them. The [N]egro has been too long associated with the white man not to have copied his vices as well as his virtues. But the [N]egro resents and utterly repudiates the efforts to blacken his good name by asserting that assaults upon women are peculiar to his race. The [N]egro has suffered far more from the commission of this crime against the women of his race by white men than the white race has ever suffered through his crimes. Very scant notice is taken of the matter when this is the condition of affairs. What becomes a crime deserving capital punishment when the tables are turned is a matter of small moment when the [N]egro woman is the accusing party.

But since the world has accepted this false and unjust statement, and the burden of proof has been placed upon the [N]egro to vindicate his race, he is taking steps to do so. The Anti-Lynching Bureau of the National Afro-American Council is arranging to have every lynching investigated and publish the facts to the world, as has been done in the case of Sam Hose, who was burned alive last April at Newman, Ga. The detective's report showed that Hose killed Cranford, his employer, in self-defense, and that, while a mob was organizing to hunt Hose to punish him for killing a white man, not till twenty-four hours after the murder was the charge of rape, embellished with psychological and physical impossibilities, circulated. That gave an impetus to the hunt, and the *Atlanta Constitution*'s reward of $500 keyed the mob to the necessary burning and roasting pitch. Of five hundred newspaper clippings of that horrible affair, nine-tenths of them assumed Hose's guilt—simply because his murderers said so, and because it is the fashion to believe the [N]egro peculiarly addicted to this species of crime. All the [N]egro asks is justice—a fair and impartial trial in the courts of the country. That given, he will abide the result.

But this question affects the entire American nation, and from several points of view: First, on the ground of consistency. Our watchword has been "the land of the free and the home of the brave." Brave men do not gather by thousands to torture and murder a single individual, so gagged and bound he cannot make even feeble resistance or defense. Neither do brave men or women stand by and see such things done without compunction of conscience, nor read of them without protest. Our nation has been active and outspoken in its endeavors to right the wrongs of the Armenian Christian, the Russian Jew, the Irish Home Ruler, the native women of India, the Siberian exile, and the Cuban patriot. Surely it should be the nation's duty to correct its own evils!

Second, on the ground of economy. To those who fail to be convinced from any other point of view touching this momentous question, a consideration of the economic phase might not be amiss. It is generally known that mobs in Louisiana, Colorado, Wyoming, and other States have lynched subjects of other countries. When their different governments demanded satisfaction, our country was forced to confess her inability to protect said subjects in the several States because of our State-rights doctrines, or in turn demand punishment of the lynchers.

This confession, while humiliating in the extreme, was not satisfactory; and, while the United States cannot protect, she can pay. This she has done, and it is certain will have to do again in the case of the recent lynching of Italians in Louisiana. The United States already has paid in indemnities for lynching nearly a half million dollars, as follows:

Paid China for Rock Springs (Wyo.) massacre	$ 147,748.74
Paid China for outrages on Pacific Coast	276,619.75
Paid Italy for massacre of Italian prisoners at New Orleans	24,330.90
Paid Italy for lynchings at Walsenburg, Col.	10,000.00
Paid Great Britain for outrages on James Bainand Frederick Dawson	2,800.00
GRAND TOTAL	**$ 461,499.99**

Third, for the honor of Anglo-Saxon civilization. No scoffer at our boasted American civilization could say anything more harsh of it than does the American white man himself who says he is unable to protect the honor of his women without resort to such brutal, inhuman, and degrading exhibitions as characterize "lynching bees." The cannibals of the South Sea Islands roast human beings alive to satisfy hunger. The red Indian of the Western plains tied his prisoner to the stake, tortured him, and danced in fiendish glee while his victim writhed in the flames. His savage, untutored mind suggested no better way than that of wreaking vengeance upon those who had wronged him. These people knew nothing about Christianity and did not profess to follow its teachings; but such primary laws as they had they lived up to. No nation, savage or civilized, save only the United States of America, has confessed its inability to protect its women save by hanging, shooting, and burning alleged offenders.

Finally, for love of country. No American travels abroad without blushing for shame for his country on this subject. And whatever the excuse that passes current in the United States, it avails nothing abroad.

With all the powers of government in control; with all laws made by white men, administered by white judges, jurors, prosecuting attorneys, and sheriffs; with every office of the executive department filled by white men—no excuse can be offered for exchanging the orderly administration of justice for barbarous lynchings and "unwritten laws." Our country should be placed speedily above the plane of confessing herself a failure at self-government. This cannot be until Americans of every section, of broadest patriotism and best and wisest citizenship, not only see the defect in our country's armor but take the necessary steps to remedy it. Although lynchings have steadily increased in number and barbarity during the last twenty years, there has been no single effort put forth by the many moral and philanthropic forces of the country to put a stop to this wholesale slaughter. Indeed, the silence and seeming condonation grow more marked as the years go by.

A few months ago the conscience of this country was shocked because, after a two-weeks trial, a French judicial tribunal pronounced Captain Dreyfus guilty. And yet, in our own land and under our own flag, the writer can give day and detail of one thousand men, women, and children who during the last six years were put to death without trial before any tribunal on earth. Humiliating indeed, but altogether unanswerable, was the reply of the French press to our protest: "Stop your lynchings at home before you send your protests abroad."

DISCUSSION QUESTIONS

1. Wells-Barnett uses male-specific terms in writing about lynching. Although this could have been a linguistic convention of the time, she seems to explain the practice of lynching in this gender-specific way. How would cultural norms of masculinity have affected this disturbing practice?

2. The sociological theorist Émile Durkheim argued that public executions reaffirm the collective consciousness of dominant groups in society. Would Wells-Barnett agree?

3. What does Wells-Barnett mean by "unwritten laws," and how does this show the power of group conformity, even in the absence of formal codes of conduct?

6

The Self

GEORGE HERBERT MEAD

In our statement of the development of intelligence we have already suggested that the language process is essential for the development of the self. The self has a character which is different from that of the physiological organism proper. The self is something which has a development; it is not initially there, at birth, but arises in the process of social experience and activity, that is, develops in the given individual as a result of his relations to that process as a whole and to other individuals within that process....

We can distinguish very definitely between the self and the body. The body can be there and can operate in a very intelligent fashion without there being a self involved in the experience. The self has the characteristic that it is an object to itself, and that characteristic distinguishes it from other objects and from the body. It is perfectly true that the eye can see the foot, but it does not see the body as a whole. We cannot see our backs; we can feel certain portions of them, if we are agile, but we cannot get an experience of our whole body. There are, of course, experiences which are somewhat vague and difficult of location, but the bodily experiences are for us organized about a self. The foot and hand belong to the self. We can see our feet, especially if we look at them from the wrong end of an opera glass, as strange things which we have difficulty in recognizing as our own. The parts of the body are quite distinguishable from the self. We can lose parts of the body without any serious invasion of the self. The mere ability to experience different parts of the body is not different from the experience of a table. The table presents a different feel from what the hand does when one hand feels another, but it is an experience of something with which we come definitely into contact. The body does not experience itself as a whole, in the sense in which the self in some way enters into the experience of the self.

SOURCE: "The Self," from *Mind, Self, and Society* by George Herbert Mead, Copyright 1934. Reprinted by permission of the University of Chicago Press.

It is the characteristic of the self as an object to itself that I want to bring out. This characteristic is represented in the word "self," which is a reflexive, and indicates that which can be both subject and object. This type of object is essentially different from other objects, and in the past it has been distinguished as conscious, a term which indicates an experience with, an experience of, one's self. It was assumed that consciousness in some way carried this capacity of being an object to itself. In giving a behavioristic statement of consciousness we have to look for some sort of experience in which the physical organism can become an object to itself.[1]

When one is running to get away from someone who is chasing him, he is entirely occupied in this action, and his experience may be swallowed up in the objects about him, so that he has, at the time being, no consciousness of self at all. We must be, of course, very completely occupied to have that take place, but we can, I think, recognize that sort of a possible experience in which the self does not enter. We can, perhaps, get some light on that situation through those experiences in which in very intense action there appear in the experience of the individual, back of this intense action, memories and anticipations. Tolstoi as an officer in the war gives an account of having pictures of his past experience in the midst of his most intense action. There are also the pictures that flash into a person's mind when he is drowning. In such instances there is a contrast between an experience that is absolutely wound up in outside activity in which the self as an object does not enter, and an activity of memory and imagination in which the self is the principal object. The self is then entirely distinguishable from an organism that is surrounded by things and acts with reference to things, including parts of its own body. These latter may be objects like other objects, but they are just objects out there in the field, and they do not involve a self that is an object to the organism. This is, I think, frequently overlooked. It is that fact which makes our anthropomorphic reconstructions of animal life so fallacious. How can an individual get outside himself (experientially) in such a way as to become an object to himself? This is the essential psychological problem of selfhood or of self-consciousness; and its solution is to be found by referring to the process of social conduct or activity in which the given person or individual is implicated. The apparatus of reason would not be complete unless it swept itself into its own analysis of the field of experience; or unless the individual brought himself into the same experiential field as that of the other individual selves in relation to whom he acts in any given social situation. Reason cannot become impersonal unless it takes an objective, nonaffective attitude toward itself; otherwise we have just consciousness, not *self*-consciousness. And it is necessary to rational conduct that the individual should thus take an objective, impersonal attitude toward himself, that he should become an object to himself. For the individual organism is obviously an essential and important fact or constituent element of the empirical situation in which it acts; and without taking objective account of itself as such, it cannot act intelligently, or rationally.

The individual experiences himself as such, not directly, but only indirectly, from the particular standpoints of other individual members of the same social group, or from the generalized standpoint of the social group as a whole to which he belongs. For he enters his own experience as a self or individual, not directly or immediately, not by becoming a subject to himself, but only insofar as he first becomes an object to himself just as other individuals are objects to him or in his experience; and he becomes an object to himself only by taking the attitudes of other individuals toward himself within a social environment or context of experience and behavior in which both he and they are involved.

The importance of what we term "communication" lies in the fact that it provides a form of behavior in which the organism or the individual may become an object to himself. It is that sort of communication which we have been discussing—not communication in the sense of the cluck of the hen to the chickens, or the bark of a wolf to the pack, or the lowing of a cow, but communication in the sense of significant symbols, communication which is directed not only to others but also to the individual himself. So far as that type of communication is a part of behavior it at least introduces a self. Of course, one may hear without listening; one may see things that he does not realize; do things that he is not really aware of. But it is where one does respond to that which he addresses to another and where that response of his own becomes a part of his conduct, where he not only hears himself but responds to himself, talks and replies to himself as truly as the other person replies to him, that we have behavior in which the individuals become objects to themselves....

The self, as that which can be an object to itself, is essentially a social structure, and it arises in social experience. After a self has arisen, it in a certain sense provides for itself its social experiences, and so we can conceive of an absolutely solitary self. But it is impossible to conceive of a self arising outside of social experience. When it has arisen we can think of a person in solitary confinement for the rest of his life, but who still has himself as a companion, and is able to think and to converse with himself as he had communicated with others. That process to which I have just referred, of responding to one's self as another responds to it, taking part in one's own conversation with others, being aware of what one is saying and using that awareness of what one is saying to determine what one is going to say thereafter—that is a process with which we are all familiar. We are continually following up our own address to other persons by an understanding of what we are saying, and using that understanding in the direction of our continued speech. We are finding out what we are going to say, what we are going to do, by saying and doing, and in the process we are continually controlling the process itself. In the conversation of gestures what we say calls out a certain response in another and that in turn changes our own action, so that we shift from what we started to do because of the reply the other makes. The conversation of gestures is the beginning of communication. The individual comes to carry on a conversation of gestures with himself. He says something, and that calls out a certain reply in himself which makes him change what he was going to say. One starts to say something, we will presume an unpleasant something, but when he starts to say it he realizes it is cruel. The effect on himself of what he is saying

checks him; there is here a conversation of gestures between the individual and himself. We mean by significant speech that the action is one that affects the individual himself, and that the effect upon the individual himself is part of the intelligent carrying-out of the conversation with others. Now we, so to speak, amputate that social phase and dispense with it for the time being, so that one is talking to one's self as one would talk to another person.[2]

This process of abstraction cannot be carried on indefinitely. One inevitably seeks an audience, has to pour himself out to somebody. In reflective intelligence one thinks to act, and to act solely so that this action remains a part of a social process. Thinking becomes preparatory to social action. The very process of thinking is, of course, simply an inner conversation that goes on, but it is a conversation of gestures which in its completion implies the expression of that which one thinks to an audience. One separates the significance of what he is saying to others from the actual speech and gets it ready before saying it. He thinks it out, and perhaps writes it in the form of a book; but it is still a part of social intercourse in which one is addressing other persons and at the same time addressing one's self, and in which one controls the address to other persons by the response made to one's own gesture. That the person should be responding to himself is necessary to the self, and it is this sort of social conduct which provides behavior within which that self appears. I know of no other form of behavior than the linguistic in which the individual is an object to himself, and, so far as I can see, the individual is not a self in the reflexive sense unless he is an object to himself. It is this fact that gives a critical importance to communication, since this is a type of behavior in which the individual does so respond to himself.

We realize in everyday conduct and experience that an individual does not mean a great deal of what he is doing and saying. We frequently say that such an individual is not himself. We come away from an interview with a realization that we have left out important things, that there are parts of the self that did not get into what was said. What determines the amount of the self that gets into communication is the social experience itself. Of course, a good deal of the self does not need to get expression. We carry on a whole series of different relationships to different people. We are one thing to one man and another thing to another. There are parts of the self which exist only for the self in relationship to itself. We divide ourselves up in all sorts of different selves with reference to our acquaintances. We discuss politics with one and religion with another. There are all sorts of different selves answering to all sorts of different social reactions. It is the social process itself that is responsible for the appearance of the self; it is not there as a self apart from this type of experience.

A multiple personality is in a certain sense normal, as I have just pointed out....

The unity and structure of the complete self reflects the unity and structure of the social process as a whole; and each of the elementary selves of which it is composed reflects the unity and structure of one of the various aspects of that process in which the individual is implicated. In other words, the various elementary selves which constitute, or are organized into, a complete self are the various aspects of the structure of that complete self answering to the various aspects of the structure of the social process as a whole; the structure of the complete self is thus a reflection of the complete social process. The organization and unification of a social group is identical with the organization and unification of any one of the selves arising within the social process in which that group is engaged, or which it is carrying on.[3]...

Another set of background factors in the genesis of the self is represented in the activities of play and the game.... We find in children ... imaginary companions which a good many children produce in their own experience. They organize in this way the responses which they call out in other persons and call out also in themselves. Of course, this playing with an imaginary companion is only a peculiarly interesting phase of ordinary play. Play in this sense, especially the stage which precedes the organized games, is a play at something. A child plays at being a mother, at being a teacher, at being a policeman; that is, it is taking different roles, as we say. We have something that suggests this in what we call the play

of animals: A cat will play with her kittens, and dogs play with each other. Two dogs playing with each other will attack and defend, in a process which if carried through would amount to an actual fight. There is a combination of responses which checks the depth of the bite. But we do not have in such a situation the dogs taking a definite role in the sense that a child deliberately takes the role of another. This tendency on the part of children is what we are working with in the kindergarten where the roles which the children assume are made the basis for training. When a child does assume a role he has in himself the stimuli which call out that particular response or group of responses. He may, of course, run away when he is chased, as the dog does, or he may turn around and strike back just as the dog does in his play. But that is not the same as playing at something. Children get together to "play Indian." This means that the child has a certain set of stimuli that call out in itself the responses that they would call out in others, and which answer to an Indian. In the play period the child utilizes his own responses to these stimuli which he makes use of in building a self. The response which he has a tendency to make to these stimuli organizes them. He plays that he is, for instance, offering himself something, and he buys it; he gives a letter to himself and takes it away; he addresses himself as a parent, as a teacher; he arrests himself as a policeman. He has a set of stimuli which call out in himself the sort of responses they call out in others. He takes this group of responses and organizes them into a certain whole. Such is the simplest form of being another to one's self. It involves a temporal situation. The child says something in one character and responds in another character, and then his responding in another character is a stimulus to himself in the first character, and so the conversation goes on. A certain organized structure arises in him and in his other which replies to it, and these carry on the conversation of gestures between themselves.

If we contrast play with the situation in an organized game, we note the essential difference that the child who plays in a game must be ready to take the attitude of everyone else involved in that game, and that these different roles must have definite relationship to each other. Taking a very simple game such as hide-and-seek, everyone with the exception of the one who is hiding is a person who is hunting. A child does not require more than the person who is hunted and the one who is hunting. If a child is playing in the first sense he just goes on playing, but there is no basic organization gained. In that early stage he passes from one to another just as a whim takes him. But in a game where a number of individuals are involved, then the child taking one role must be ready to take the role of everyone else. If he gets in a ball nine he must have the responses of each position involved in his own position. He must know what everyone else is going to do in order to carry out his own play. He has to take all of these roles. They do not all have to be present in consciousness at the same time, but at some moments he has to have three or four individuals present in his own attitude, such as the one who is going to throw the ball, the one who is going to catch it, and so on. These responses must be, in some degree, present in his own makeup. In the game, then, there is a set of responses of such others so organized that the attitude of one calls out the appropriate attitudes of the other.

This organization is put in the form of the rules of the game. Children take a great interest in rules. They make rules on the spot in order to help themselves out of difficulties. Part of the enjoyment of the game is to get these rules. Now, the rules are the set of responses which a particular attitude calls out. You can demand a certain response in others if you take a certain attitude. These responses are all in yourself as well. There you get an organized set of such responses as that to which I have referred, which is something more elaborate than the roles found in play. Here there is just a set of responses that follow on each other indefinitely. At such a stage we speak of a child as not yet having a fully developed self. The child responds in a fairly intelligent fashion to the immediate stimuli that come to him, but they are not organized. He does not organize his life as we would like to have him do, namely, as a whole. There is just a set of responses of the type of play. The child reacts to a certain stimulus, and the reaction is in himself that is called out in others, but he is

not a whole self. In his game he has to have an organization of these roles; otherwise he cannot play the game. The game represents the passage in the life of the child from taking the role of others in play to the organized part that is essential to self-consciousness in the full sense of the term....

The fundamental difference between the game and play is that in the former the child must have the attitude of all the others involved in that game. The attitudes of the other players which the participant assumes organize into a sort of unit, and it is that organization which controls the response of the individual. The illustration used was of a person playing baseball. Each one of his own acts is determined by his assumption of the action of the others who are playing the game. What he does is controlled by his being everyone else on that team, at least insofar as those attitudes affect his own particular response. We get then an "other" which is an organization of the attitudes of those involved in the same process.

The organized community or social group which gives to the individual his unity of self may be called "the generalized other." The attitude of the generalized other is the attitude of the whole community.[4] Thus, for example, in the case of such a social group as a ball team, the team is the generalized other insofar as it enters—as an organized process or social activity—into the experience of any one of the individual members of it.

If the given human individual is to develop a self in the fullest sense, it is not sufficient for him merely to take the attitudes of other human individuals toward himself and toward one another within the human social process, and to bring that social process as a whole into his individual experience merely in these terms: He must also, in the same way that he takes the attitudes of other individuals toward himself and toward one another, take their attitudes toward the various phases or aspects of the common social activity or set of social undertakings in which, as members of an organized society or social group, they are all engaged; and he must then, by generalizing these individual attitudes of that organized society or social group itself, as a whole, act toward different social projects which at any given time it is carrying out, or toward the various larger phases of the general social process which constitutes its life and of which these projects are specific manifestations. This getting of the broad activities of any given social whole or organized society as such within the experiential field of any one of the individuals involved or included in that whole is, in other words, the essential basis and prerequisite of the fullest development of that individual's self: Only insofar as he takes the attitudes of the organized social group to which he belongs toward the organized, cooperative social activity or set of such activities in which that group as such is engaged, does he develop a complete self or possess the sort of complete self he has developed. And on the other hand, the complex cooperative processes and activities and institutional functionings of organized human society are also possible only insofar as every individual involved in them or belonging to that society can take the general attitudes of all other such individuals with reference to these processes and activities and institutional functionings, and to the organized social whole of experiential relations and interactions thereby constituted—and can direct his own behavior accordingly.

It is in the form of the generalized other that the social process influences the behavior of the individuals involved in it and carrying it on, i.e., that the community exercises control over the conduct of its individual members; for it is in this form that the social process or community enters as a determining factor into the individual's thinking. In abstract thought the individual takes the attitude of the generalized other[5] toward himself, without reference to its expression in any particular other individuals; and in concrete thought he takes that attitude insofar as it expresses in the attitudes toward his behavior of those other individuals with whom he is involved in the given social situation or act. But only by taking the attitude of the generalized other toward himself, in one or another of these ways, can he think at all; for only thus can thinking—or the internalized conversation of gestures which constitutes thinking—occur. And only through the taking by individuals of the attitude or attitudes of the generalized other toward themselves is the existence of a universe of

discourse, as that system of common or social meanings which thinking presupposes at its context, rendered possible....

I have pointed out, then, that there are two general stages in the full development of the self. At the first of these stages, the individual's self is considered simply by an organization of the particular attitudes of other individuals toward himself and toward one another in the specific social acts in which he participates with them. But at the second stage in the full development of the individual's self that self is constituted not only by an organization of these particular individual attitudes, but also by an organization of the social attitudes of the generalized other or the social group as a whole to which he belongs.... So the self reaches its full development by organizing these individual attitudes of others into the organized social or group attitudes, and by thus becoming an individual reflection of the general systematic pattern of social or group behavior in which it and the others are all involved—a pattern which enters as a whole into the individual's experience in terms of these organized group attitudes which, through the mechanism of his central nervous system, he takes toward himself, just as he takes the individual attitudes of others....

A person is a personality because he belongs to a community, because he takes over the institutions of that community into his own conduct. He takes its language as a medium by which he gets his personality, and then through a process of taking the different roles that all the others furnish he comes to get the attitude of the members of the community. Such, in a certain sense, is the structure of a man's personality. There are certain common responses which each individual has toward certain common things, and insofar as those common responses are awakened in the individual when he is affecting other persons he arouses his own self. The structure, then, on which the self is built is this response which is common to all, for one has to be a member of a community to be a self. Such responses are abstract attitudes, but they constitute just what we term a man's character. They give him what we term his principles, the acknowledged attitudes of all members of the community toward what are the values of that community. He is putting himself in the place of the generalized other, which represents the organized responses of all the members of the group. It is that which guides conduct controlled by principles, and a person who has such an organized group of responses is a man whom we say has character, in the moral sense....

I have so far emphasized what I have called the structures upon which the self is constructed, the framework of the self, as it were. Of course we are not only what is common to all: Each one of the selves is different from everyone else; but there has to be such a common structure as I have sketched in order that we may be members of a community at all. We cannot be ourselves unless we are also members in whom there is a community of attitudes which control the attitudes of all. We cannot have rights unless we have common attitudes. That which we have acquired as self-conscious persons makes us such members of society and gives us selves. Selves can only exist in definite relationships to other selves. No hard-and-fast line can be drawn between our own selves and the selves of others, since our own selves exist and enter as such into our experience only insofar as the selves of others exist and enter as such into our experience also. The individual possesses a self only in relation to the selves of the other members of his social group; and the structure of his self expresses or reflects the general behavior pattern of this social group to which he belongs, just as does the structure of the self of every other individual belonging to this social group.

NOTES

1. Man's behavior is such in his social group that he is able to become an object to himself, a fact which constitutes him a more advanced product of evolutionary development than are the lower animals. Fundamentally it is this social fact—and not his alleged possession of a soul or mind with which he, as an individual, has been mysteriously and supernaturally endowed, and with which the lower animals have not been endowed—that differentiates him from them.

2. It is generally recognized that the specifically social expressions of intelligence, or the exercise of what is often called "social intelligence," depend upon the given individual's ability to take the roles of, or "put himself in the place of," the other individuals implicated with him in given social situations; and upon his consequent sensitivity to their attitudes toward himself and toward one another. These specifically social expressions of intelligence, of course, acquire unique significance in terms of our view that the whole nature of intelligence is social to the very core—that this putting of one's self in the places of others, this taking by one's self of their roles or attitudes, is not merely one of the various aspects or expressions of intelligence or intelligent behavior, but is the very essence of its character. Spearman's "X factor" in intelligence—the unknown factor which, according to him, intelligence contains—is simply (if our social theory of intelligence is correct) this ability of the intelligent individual to take the attitude of the other, or the attitudes of others, thus realizing the significations or grasping the meanings of the symbols or gestures in terms of which thinking proceeds; and thus being able to carry on with himself the internal conversation with these symbols or gestures which thinking involves.

3. The unity of the mind is not identical with the unity of the self. The unity of the self is constituted by the unity of the entire relational pattern of social behavior and experience in which the individual is implicated, and which is reflected in the structure of the self; but many of the aspects or features of this entire pattern do not enter into consciousness, so that the unity of the mind is in a sense an abstraction from the more inclusive unity of the self.

4. It is possible for inanimate objects, no less than for other human organisms, to form parts of the generalized and organized—the completely socialized—other for any given human individual, insofar as he responds to such objects socially or in a social fashion (by means of the mechanism of thought, the internalized conversation of gestures). Any thing—any object or set of objects, whether animate or

inanimate, human or animal, or merely physical—toward which he acts, or to which he responds, socially, is an element in what for him is the generalized other, by taking the attitudes of which toward himself he becomes conscious of himself as an object or individual, and thus develops a self or personality. Thus, for example, the cult, in its primitive form, is merely the social embodiment of the relation between the given social group or community and its physical environment—an organized social means, adopted by the individual members of that group or community, of entering into social relations with that environment, or (in a sense) of carrying on conversations with it; and in this way that environment becomes part of the total generalized other for each of the individual members of the given social group or community.

5.   We have said that the internal conversation of the individual with himself in terms of words or significant gestures—the conversation which constitutes the process or activity of thinking—is carried on by the individual from the standpoint of the "generalized other." And the more abstract that conversation is, the more abstract thinking happens to be, the further removed is the generalized other from any connection with particular individuals. It is especially in abstract thinking, that is to say, that the conversation involved is carried on by the individual with the generalized other, rather than with any particular individuals. Thus it is, for example, that abstract concepts are concepts stated in terms of the attitudes of the entire social group or community; they are stated on the basis of the individual's consciousness of the attitudes of the generalized other toward them, as a result of his taking these attitudes of the generalized other and then responding to them. And thus it is also that abstract propositions are stated in a form which anyone—any other intelligent individual—will accept.

DISCUSSION QUESTIONS

1.   George Herbert Mead shows how children's play and games are fundamental in learning the "self." What play and games influenced the formation of your self-concept? How does this vary for different social groups (i.e., groups distinguished by gender, race, class, age, etc.)?

2.   What role does reflection have in the formation of the self? Why, according to Mead, is this fundamentally a social process?

3.   How does Mead distinguish the self and the body? Taking his analysis further, how do people's reflections on the body influence the formation of self? What other social factors are involved in this process, and what do they tell you about the relationship between the individual and society?

INTERNET RESOURCES

Suggested Web URLs for Further Study

http://www.brocku.ca/MeadProject/
The only site you need for George Herbert Mead contains related works, analyses, research, and discussion forums.

Virtual Society

For further information on this subject, including links to relevant websites, go to the Wadsworth Sociology homepage at *http://www.cengage.com/sociology*.

7

The Presentation of Self

ERVING GOFFMAN

When an individual enters the presence of others, they commonly seek to acquire information about him or to bring into play information about him already possessed. They will be interested in his general socioeconomic status, his conception of self, his attitude toward them, his competence, his trustworthiness, etc. Although some of this information seems to be sought almost as an end in itself, there are usually quite practical reasons for acquiring it. Information about the individual helps to define the situation, enabling others to know in advance what he will expect of them and what they may expect of him. Informed in these ways, the others will know how best to act in order to call forth a desired response from him.

For those present, many sources of information become accessible and many carriers (or "signvehicles") become available for conveying this information. If unacquainted with the individual, observers can glean clues from his conduct and appearance which allow them to apply their previous experience with individuals roughly similar to the one before them or, more important, to apply untested stereotypes to him. They can also assume from past experience that only individuals of a particular kind are likely to be found in a given social setting. They can rely on what the individual says about himself or on documentary evidence he provides as to who and what he is. If they know, or know of, the individual by virtue of experience prior to the interaction, they can rely on assumptions as to the persistence and generality of psychological traits as a means of predicting his present and future behavior.

However, during the period in which the individual is in the immediate presence of the others, few events may occur which directly provide the others with the conclusive information they will need if they are to direct wisely their own activity. Many crucial facts lie beyond the time and place of interaction or lie

SOURCE: From *The Presentation of Self in Everyday Life* by Erving Goffman, © 1959 by Erving Goffman. Used by permission of Doubleday, a division of Random House, Inc.

concealed within it. For example, the "true" or "real" attitudes, beliefs, and emotions of the individual can be ascertained only indirectly, through his avowals or through what appears to be involuntary expressive behavior. Similarly, if the individual offers the others a product or service, they will often find that during the interaction there will be no time and place immediately available for eating the pudding that the proof can be found in. They will be forced to accept some events as conventional or natural signs of something not directly available to the senses. In Ichheiser's terms,[1] the individual will have to act so that he intentionally or unintentionally *expresses* himself, and the others will in turn have to be *impressed* in some way by him.

The expressiveness of the individual (and therefore his capacity to give impressions) appears to involve two radically different kinds of sign activity: the expression that he *gives*, and the expression that he *gives off*. The first involves verbal symbols or their substitutes which he uses admittedly and solely to convey the information that he and the others are known to attach to these symbols. This is communication in the traditional and narrow sense. The second involves a wide range of action that others can treat as symptomatic of the actor, the expectation being that the action was performed for reasons other than the information conveyed in this way. As we shall have to see, this distinction has an only initial validity. The individual does of course intentionally convey misinformation by means of both of these types of communication, the first deceit, the second feigning....

Let us now turn from the others to the point of view of the individual who presents himself before them. He may wish them to think highly of him, or to think that he thinks highly of them, or to perceive how in fact he feels toward them, or to obtain no clear-cut impression; he may wish to ensure sufficient harmony so that the interaction can be sustained, or to defraud, get rid of, confuse, mislead, antagonize, or insult them. Regardless of the particular objective which the individual has in mind and of his motive for having this objective, it will be in his interests to control the conduct of the others, especially their responsive treatment of him. This control is achieved largely by influencing the definition of the situation

which the others come to formulate, and he can influence this definition by expressing himself in such a way as to give them the kind of impression that will lead them to act voluntarily in accordance with his own plan. Thus, when an individual appears in the presence of others, there will usually be some reason for him to mobilize his activity so that it will convey an impression to others which it is in his interests to convey. Since a girl's dormitory mates will glean evidence of her popularity from the calls she receives on the phone, we can suspect that some girls will arrange for calls to be made, and Willard Waller's finding can be anticipated:

> It has been reported by many observers that a girl who is called to the telephone in the dormitories will often allow herself to be called several times, in order to give all the other girls ample opportunity to hear her paged.[2]

Of the two kinds of communication—expressions given and expressions given off—this report will be primarily concerned with the latter, with the more theatrical and contextual kind, the nonverbal, presumably unintentional kind, whether this communication be purposely engineered or not. As an example of what we must try to examine, I would like to cite at length a novelistic incident in which Preedy, a vacationing Englishman, makes his first appearance on the beach of his summer hotel in Spain:

> But in any case he took care to avoid catching anyone's eye. First of all, he had to make it clear to those potential companions of his holiday that they were of no concern to him whatsoever. He stared through them, round them, over them—eyes lost in space. The beach might have been empty. If by chance a ball was thrown his way, he looked surprised; then let a smile of amusement lighten his face (Kindly Preedy), looked round dazed to see that there *were* people on the beach, tossed it back with a smile to himself and not a smile at the people, and then resumed carelessly his nonchalant survey of space.

But it was time to institute a little parade, the parade of the Ideal Preedy. By devious handlings he gave any who wanted to look a chance to see the title of his book—a Spanish translation of Homer, classic thus, but not daring, cosmopolitan too—and then gathered together his beach-wrap and bag into a neat sand-resistant pile (Methodical and Sensible Preedy), rose slowly to stretch at ease his huge frame (Big-Cat Preedy), and tossed aside his sandals (Carefree Preedy, after all).

The marriage of Preedy and the sea! There were alternative rituals. The first involved the stroll that turns into a run and a dive straight into the water, thereafter smoothing into a strong splashless crawl towards the horizon. But of course not really to the horizon. Quite suddenly he would turn on to his back and thrash great white splashes with his legs, somehow thus showing that he could have swum further had he wanted to, and then would stand up a quarter out of water for all to see who it was.

The alternative course was simpler, it avoided the cold-water shock and it avoided the risk of appearing too high-spirited. The point was to appear to be so used to the sea, the Mediterranean, and this particular beach, that one might as well be in the sea as out of it. It involved a slow stroll down and into the edge of the water—not even noticing his toes were wet, land and water all the same to *him!*—with his eyes up at the sky gravely surveying portents, invisible to others, of the weather (Local Fisherman Preedy).[3]

The novelist means us to see that Preedy is improperly concerned with the extensive impressions he feels his sheer bodily action is giving off to those around him. We can malign Preedy further by assuming that he has acted merely in order to give a particular impression, that this is a false impression, and that the others present receive either no impression at all, or, worse still, the impression

that Preedy is affectedly trying to cause them to receive this particular impression. But the important point for us here is that the kind of impression Preedy thinks he is making is in fact the kind of impression that others correctly and incorrectly glean from someone in their midst....

There is one aspect of the others' response that bears special comment here. Knowing that the individual is likely to present himself in a light that is favorable to him, the others may divide what they witness into two parts; a part that is relatively easy for the individual to manipulate at will, being chiefly his verbal assertions, and a part in regard to which he seems to have little concern or control, being chiefly derived from the expressions he gives off. The others may then use what are considered to be the ungovernable aspects of his expressive behavior as a check upon the validity of what is conveyed by the governable aspects. In this a fundamental asymmetry is demonstrated in the communication process, the individual presumably being aware of only one stream of his communication, the witnesses of this stream and one other. For example, in Shetland Isle one crofter's wife, in serving native dishes to a visitor from the mainland of Britain, would listen with a polite smile to his polite claims of liking what he was eating; at the same time she would take note of the rather rapidity with which the visitor lifted his fork or spoon to his mouth, the eagerness with which he passed food into his mouth, and the gusto expressed in chewing the food, using these signs as a check on the stated feelings of the eater. The same woman, in order to discover what one acquaintance (A) "actually" thought of another acquaintance (B), would wait until B was in the presence of A but engaged in conversation with still another person (C). She would then covertly examine the facial expressions of A as he regarded B in conversation with C. Not being in conversation with B, and not being directly observed by him, A would sometimes relax usual constraints and tactful deceptions, and freely express what he was "actually" feeling about B. This Shetlander, in short, would observe the unobserved observer.

Now given the fact that others are likely to check up on the more controllable aspects of behavior by means of the less controllable, one can expect that sometimes the individual will try to exploit this very possibility, guiding the impression he makes through behavior felt to be reliably informing.[4] For example, in gaining admission to a tight social circle, the participant observer may not only wear an accepting look while listening to an informant, but may also be careful to wear the same look when observing the informant talking to others; observers of the observer will then not as easily discover where he actually stands. A specific illustration may be cited from Shetland Isle. When a neighbor dropped in to have a cup of tea, he would ordinarily wear at least a hint of an expectant warm smile as he passed through the door into the cottage. Since lack of physical obstructions outside the cottage and lack of light within it usually made it possible to observe the visitor unobserved as he approached the house, islanders sometimes took pleasure in watching the visitor drop whatever expression he was manifesting and replace it with a sociable one just before reaching the door. However, some visitors, in appreciating that this examination was occurring, would blindly adopt a social face a long distance from the house, thus ensuring the projection of a constant image.

This kind of control upon the part of the individual reinstates the symmetry of the communication process, and sets the stage for a kind of information game—a potentially infinite cycle of concealment, discovery, false revelation, and rediscovery. It should be added that since the others are likely to be relatively unsuspicious of the presumably unguided aspects of the individual's conduct, he can gain much by controlling it. The others of course may sense that the individual is manipulating the presumably spontaneous aspects of his behavior, and seek in this very act of manipulation some shading of conduct that the individual has not managed to control. This again provides a check upon the individual's behavior, this time his presumably uncalculated behavior, thus reestablishing the asymmetry of the communication process. Here I would like only to add the suggestion that the arts of piercing an individual's effort at calculated unintentionality seem better developed than our capacity to manipulate our own behavior,

so that regardless of how many steps have occurred in the information game, the witness is likely to have the advantage over the actor, and the initial asymmetry of the communication process is likely to be retained....

In everyday life, of course, there is a clear understanding that first impressions are important. Thus, the work adjustment of those in service occupations will often hinge upon a capacity to seize and hold the initiative in the service relation, a capacity that will require subtle aggressiveness on the part of the server when he is of lower socioeconomic status than his client. W. F. Whyte suggests the waitress as an example:

> The first point that stands out is that the waitress who bears up under pressure does not simply respond to her customers. She acts with some skill to control their behavior. The first question to ask when we look at the customer relationship is, "Does the waitress get the jump on the customer, or does the customer get the jump on the waitress?" The skilled waitress realizes the crucial nature of this question
>
> The skilled waitress tackles the customer with confidence and without hesitation. For example, she may find that a new customer has seated himself before she could clear off the dirty dishes and change the cloth. He is now leaning on the table studying the menu. She greets him, says, "May I change the cover, please?" and, without waiting for an answer, takes his menu away from him so that he moves back from the table, and she goes about her work. The relationship is handled politely but firmly, and there is never any question as to who is in charge.[5]

When the interaction that is initiated by "first impressions" is itself merely the initial interaction in an extended series of interactions involving the same participants, we speak of "getting off on the right foot" and feel that it is crucial that we do so.

Thus, one learns that some teachers take the following view:

> You can't ever let them get the upper hand on you or you're through. So I start out tough. The first day I get a new class in, I let them know who's boss.... You've got to start off tough, then you can ease up as you go along. If you start out easy-going, when you try to get tough, they'll just look at you and laugh.[6]

... In stressing the fact that the initial definition of the situation projected by an individual tends to provide a plan for the cooperative activity that follows—in stressing this action point of view—we must not overlook the crucial fact that any projected definition of the situation also has a distinctive moral character. It is this moral character of projections that will chiefly concern us in this report. Society is organized on the principle that any individual who possesses certain social characteristics has a moral right to expect that others will value and treat him in an appropriate way. Connected with this principle is a second, namely that an individual who implicitly or explicitly signifies that he has certain social characteristics ought in fact to be what he claims he is. In consequence, when an individual projects a definition of the situation and thereby makes an implicit or explicit claim to be a person of a particular kind, he automatically exerts a moral demand upon the others, obliging them to value and treat him in the manner that persons of his kind have a right to expect. He also implicitly foregoes all claims to be things he does not appear to be[7] and hence foregoes the treatment that would be appropriate for such individuals. The others find, then, that the individual has informed them as to what is and as to what they *ought* to see as the "is."

One cannot judge the importance of definitional disruptions by the frequency with which they occur, for apparently they would occur more frequently were not constant precautions taken. We find that preventive practices are constantly employed to avoid these embarrassments and that corrective practices are constantly employed to compensate for discrediting occurrences that have not been successfully avoided. When the individual employs these strategies and tactics to protect his own projections, we may refer to them as "defensive practices"; when a participant employs them to save the definition of the situation projected by another, we speak of "protective practices" or "tact." Together, defensive and protective practices comprise the techniques employed to safeguard the impression fostered by an individual during his presence before others. It should be added that while we may be ready to see that no fostered impression would survive if defensive practices were not employed, we are less ready perhaps to see that few impressions could survive if those who received the impression did not exert tact in their reception of it.

In addition to the fact that precautions are taken to prevent disruption of projected definitions, we may also note that an intense interest in these disruptions comes to play a significant role in the social life of the group. Practical jokes and social games are played in which embarrassments which are to be taken unseriously are purposely engineered.[8] Fantasies are created in which devastating exposures occur. Anecdotes from the past—real, embroidered, or fictitious—are told and retold, detailing disruptions which occurred, almost occurred, or occurred and were admirably resolved. There seems to be no grouping which does not have a ready supply of these games, reveries, and cautionary tales, to be used as a source of humor, a catharsis for anxieties, and a sanction for inducing individuals to be modest in their claims and reasonable in their projected expectations. The individual may tell himself through dreams of getting into impossible positions. Families tell of the time a guest got his dates mixed and arrived when neither the house nor anyone in it was ready for him. Journalists tell of times when an all-too meaningful misprint occurred, and the paper's assumption of objectivity or decorum was humorously discredited. Public servants tell of times a client ridiculously misunderstood form instructions, giving answers which implied an unanticipated and bizarre definition of the situation.[9] Seamen, whose home away from home is rigorously he-man, tell stories of coming back home and inadvertently asking mother to "pass the fucking butter."[10] Diplomats tell of the time a near-sighted queen asked a republican ambassador about the health of his king.[11]

To summarize, then, I assume that when an individual appears before others he will have many motives for trying to control the impression they receive of the situation.

NOTES

1. Gustav Ichheiser, "Misunderstandings in Human Relations," supplement to *The American Journal of Sociology*, 55 (Sept. 1949), 6–7.

2. Willard Waller, "The Rating and Dating Complex," *American Sociological Review*, 2, 730.

3. William Sansom, *A Contest of Ladies* (London: Hogarth, 1956), pp. 230–31.

4. The widely read and rather sound writings of Stephen Potter are concerned in part with signs that can be engineered to give a shrewd observer the apparently incidental cues he needs to discover concealed virtues the gamesman does not in fact possess.

5. W. F. Whyte, "When Workers and Customers Meet," chap. 7, *Industry and Society*, ed. W. F. Whyte (New York: McGraw-Hill, 1946), pp. 132–33.

6. Teacher interview quoted by Howard S. Becker, "Social Class Variations in the Teacher-Pupil Relationship," *Journal of Educational Sociology*, 25, 459.

7. This role of the witness in limiting what it is the individual can be has been stressed by Existentialists, who see it as a basic threat to individual freedom. See Jean-Paul Sartre, *Being and Nothingness*, trans. Hazel E. Barnes (New York: Philosophical Library, 1956), pp. 364ff.

8. Goffman, op. cit., pp. 319–27.

9. Peter Blau, "Dynamics of Bureaucracy" (Ph.D. dissertation, Department of Sociology, Columbia University, forthcoming, University of Chicago Press), pp. 127–29.

10. Walter M. Beattie, Jr., "The Merchant Seaman" (unpublished M.A. Report, Department of Sociology, University of Chicago, 1950), p. 35.

11. Sir Frederick Ponsonby, *Recollections of Three Reigns* (New York: Dutton, 1952), p. 46.

DISCUSSION QUESTIONS

1. Think of a time when you were particularly concerned about making an impression on a group or person previously unknown to you. What did you do to give cues so you could control the other's impression of you? How did you know what "cues" to give? How does this support Erving Goffman's argument about the presentation of self?

2. Have you ever been in a situation in which your presentation of self failed? What happened? How does this illustrate Goffman's point that precautions are usually taken to prevent disruption of a projected definition?

3. Have you ever had a time when you had to give a presentation of self that you felt was "not you"? What happened? How does this illustrate Goffman's point that an individual who signifies certain social characteristics is assumed to be what he or she claims to be?

INTERNET RESOURCES

Suggested Web URLs for Further Study

*http://www.heart-intl.net/HEART/Stigma/Comp/
Celebrating%20Erving%20Goffman.htm*
This site is a post-mortem analysis of Erving Goffman
through his two best-known early works, *The
Presentation of Self in Everyday Life* and *Asylum.*

Virtual Society

For further information on this subject, including
links to relevant websites, go to the Wadsworth Sociology homepage at *http://www.cengage.com/sociology.*

8

The Normality of Crime

ÉMILE DURKHEIM

Here we are, then, in the presence of a conclusion in appearance quite paradoxical. Let us make no mistake. To classify crime among the phenomena of normal sociology is not to say merely that it is an inevitable, although regrettable phenomenon, due to the incorrigible wickedness of men; it is to affirm that it is a factor in public health, an integral part of all healthy societies. This result is, at first glance, surprising enough to have puzzled even ourselves for a long time. Once this first surprise has been overcome, however, it is not difficult to find reasons explaining this normality and at the same time confirming it.

In the first place crime is normal because a society exempt from it is utterly impossible. Crime, we have shown elsewhere, consists of an act that offends certain very strong collective sentiments. In a society in which criminal acts are no longer committed, the sentiments they offend would have to be found without exception in all individual consciousnesses, and they must be found to exist with the same degree as sentiments contrary to them. Assuming that this condition could actually be realized, crime would not thereby disappear; it would only change its form, for the very cause which would thus dry up the sources of criminality would immediately open up new ones.

Indeed, for the collective sentiments which are protected by the penal law[1] of a people at a specific moment of its history to take possession of the public conscience or for them to acquire a stronger hold where they have an insufficient grip, they must acquire an intensity greater than that which they had hitherto had. The community as a whole must experience them more vividly, for it can acquire from no other source the greater force necessary to control these individuals who formerly were the most refractory. For murderers to disappear, the horror of bloodshed must become greater in those social strata from which murderers are recruited; but, first it must become greater throughout the entire society. Moreover, the very absence of crime would directly contribute to produce this horror; because any sentiment seems much more respectable when it is always and uniformly respected.

One easily overlooks the consideration that these strong states of the common consciousness cannot be thus reinforced without reinforcing at the same time the more feeble states, whose violation previously gave birth to mere infraction of convention—since the weaker ones are only the prolongation, the attenuated form, of the stronger. Thus robbery and simple bad taste injure the same single altruistic sentiment, the respect for that which is another's. However, this same sentiment is less grievously offended by bad taste than by robbery; and since, in addition, the average consciousness has not sufficient intensity to react keenly to the bad taste, it is treated with greater tolerance. That is why the person guilty of bad taste is merely blamed, whereas the thief is punished.

But, if this sentiment grows stronger, to the point of silencing in all consciousnesses the inclination which disposes man to steal, he will become more sensitive to the offenses which, until then, touched him but lightly. He will react against them, then, with more energy; they will be the object of greater opprobrium, which will transform certain of them from the simple moral faults that they were and give them the quality of crimes. For example, improper contracts, or contracts improperly executed, which only incur public blame or civil damages, will become offenses in law.

Imagine a society of saints, a perfect cloister of exemplary individuals. Crimes, properly so called, will there be unknown; but faults which appear venial to the layman will create there the same scandal that the ordinary offense does in ordinary consciousnesses. If, then, this society has the power to judge and punish, it will define these acts as criminal and will treat them as such. For the same reason, the perfect and upright man judges his smallest failings with a severity that the majority reserve for acts more truly in the nature of an offense. Formerly, acts of violence against persons were more frequent than they are today, because respect for individual dignity was less strong. As this has increased, these crimes have become more rare; and also, many acts violating this sentiment have been introduced into the penal law which were not included there in primitive times....

NOTE

1.   Penal law is criminal law.—Ed.

DISCUSSION QUESTIONS

1.   Why is a society with no crime "utterly impossible"? Do you agree? Why or why not?

2.   Durkheim tells us that in earlier societies, acts of violence against individuals happened more often than they do today. According to Durkheim, why is this the case?

3.   What about crime is healthy for society? Explain.

4.   What is the role or function of punishment?

INTERNET RESOURCES

Suggested Web URLs for Further Study

http://durkheim.itgo.com
This is the alleged Durkheim Archive, with many other links.

http://www.emile-durkheim.com
This is another resource for Durkheim's sociological concepts and other works.

Virtual Society

For further information on this subject, including links to relevant websites, go to the Wadsworth Sociology homepage at *http://www.cengage.com/sociology*.

9

On Being Sane in Insane Places

D. L. ROSENHAN

If sanity and insanity exist, how shall we know them?

The question is neither capricious nor itself insane. However much we may be personally convinced that we can tell the normal from the abnormal, the evidence is simply not compelling. It is commonplace, for example, to read about murder trials wherein eminent psychiatrists for the defense are contradicted by equally eminent psychiatrists for the prosecution on the matter of the defendant's sanity. More generally, there are a great deal of conflicting data on the reliability, utility, and meaning of such terms as "sanity," "insanity," "mental illness," and "schizophrenia." Finally, as early as 1934, Benedict suggested that normality and abnormality are not universal. What is viewed as normal in one culture may be seen as quite aberrant in another. Thus, notions of normality and abnormality may not be quite as accurate as people believe they are.

To raise questions regarding normality and abnormality is in no way to question the fact that some behaviors are deviant or odd. Murder is deviant. So, too, are hallucinations. Nor does raising such questions deny the existence of the personal anguish that is often associated with "mental illness." Anxiety and depression exist. Psychological suffering exists. But normality and abnormality, sanity and insanity, and the diagnoses that flow from them may be less substantive than many believe them to be.

At its heart, the question of whether the sane can be distinguished from the insane (and whether degrees of insanity can be distinguished from each other) is a simple matter: do the salient characteristics that lead to diagnoses reside in the patients themselves or in the environments and contexts in which observers find them? From Bleuler, through Kretchmer, through the formulators of the recently revised *Diagnostic and Statistical Manual* of the American Psychiatric Association, the belief has been strong that patients present symptoms, that those

SOURCE: Rosenhan, D. L. "On Being Sane in Insane Places," *Science* 179 (Jan. 1973), pp. 250–257. Copyright © 1973 by American Association for the Advancement of Science. Reprinted with permission.

symptoms can be categorized, and, implicitly, that the sane are distinguishable from the insane. More recently, however, this belief has been questioned. Based in part on theoretical and anthropological considerations, but also on philosophical, legal, and therapeutic ones, the view has grown that psychological categorization of mental illness is useless at best and downright harmful, misleading, and pejorative at worst. Psychiatric diagnoses, in this view, are in the minds of the observers and are not valid summaries of characteristics displayed by the observed.

Gains can be made in deciding which of these is more nearly accurate by getting normal people (that is, people who do not have, and have never suffered, symptoms of serious psychiatric disorders) admitted to psychiatric hospitals and then determining whether they were discovered to be sane and, if so, how. If the sanity of such pseudopatients were always detected, there would be prima facie evidence that a sane individual can be distinguished from the insane context in which he is found. Normality (and presumably abnormality) is distinct enough that it can be recognized wherever it occurs, for it is carried within the person. If, on the other hand, the sanity of the pseudopatients were never discovered, serious difficulties would arise for those who support traditional modes of psychiatric diagnosis. Given that the hospital staff was not incompetent, that the pseudopatient had been behaving as sanely as he had been outside of the hospital, and that it had never been previously suggested that he belonged in a psychiatric hospital, such an unlikely outcome would support the view that psychiatric diagnosis betrays little about the patient but much about the environment in which an observer finds him.

This article describes such an experiment. Eight sane people gained secret admission to 12 different hospitals. Their diagnostic experiences constitute the data of the first part of this article; the remainder is devoted to a description of their experiences in psychiatric institutions. Too few psychiatrists and psychologists, even those who have worked in such hospitals, know what the experience is like. They rarely talk about it with former patients, perhaps

because they distrust information coming from the previously insane. Those who have worked in psychiatric hospitals are likely to have adapted so thoroughly to the settings that they are insensitive to the impact of that experience. And while there have been occasional reports of researchers who submitted themselves to psychiatric hospitalization, these researchers have commonly remained in the hospitals for short periods of time, often with the knowledge of the hospital staff. It is difficult to know the extent to which they were treated like patients or like research colleagues. Nevertheless, their reports about the inside of the psychiatric hospital have been valuable. This article extends those efforts.

PSEUDOPATIENTS
AND THEIR SETTINGS

The eight pseudopatients were a varied group. One was a psychology graduate student in his 20s. The remaining seven were older and "established." Among them were three psychologists, a pediatrician, a psychiatrist, a painter, and a housewife. Three pseudopatients were women, five were men. All of them employed pseudonyms, lest their alleged diagnoses embarrass them later. Those who were in mental health professions alleged another occupation in order to avoid the special attentions that might be accorded by staff, as a matter of courtesy or caution, to ailing colleagues. With the exception of myself (I was the first pseudopatient and my presence was known to the hospital administrator and chief psychologist and, so far as I can tell, to them alone), the presence of pseudopatients and the nature of the research program was not known to the hospital staffs.

The settings were similarly varied. In order to generalize the findings, admission into a variety of hospitals was sought. The 12 hospitals in the sample were located in five different states on the East and West coasts. Some were old and shabby, some were quite new. Some were research-oriented, others not. Some had good staff-patient ratios, others were quite understaffed. Only one was a strictly private hospital. All of the others were supported by state or federal funds or, in one instance, by university funds.

After calling the hospital for an appointment, the pseudopatient arrived at the admissions office complaining that he had been hearing voices. Asked what the voices said, he replied that they were often unclear, but as far as he could tell they said "empty," "hollow," and "thud." The voices were unfamiliar and were of the same sex as the pseudopatient. The choice of these symptoms was occasioned by their apparent similarity to existential symptoms. Such symptoms are alleged to arise from painful concerns about the perceived meaninglessness of one's life. It is as if the hallucinating person were saying, "My life is empty and hollow." The choice of these symptoms was also determined by the *absence* of a single report of existential psychoses in the literature.

Beyond alleging the symptoms and falsifying name, vocation, and employment, no further alterations of person, history, or circumstances were made. The significant events of the pseudopatient's life history were presented as they had actually occurred. Relationships with parents and siblings, with spouse and children, with people at work and in school, consistent with the aforementioned exceptions, were described as they were or had been. Frustrations and upsets were described along with joys and satisfactions. These facts are important to remember. If anything, they strongly biased the subsequent results in favor of detecting sanity, since none of their histories or current behaviors were seriously pathological in any way.

Immediately upon admission to the psychiatric ward, the pseudopatient ceased simulating *any* symptoms of abnormality. In some cases, there was a brief period of mild nervousness and anxiety, since none of the pseudopatients really believed that they would be admitted so easily. Indeed, their shared fear was that they would be immediately exposed as frauds and greatly embarrassed. Moreover, many of them had never visited a psychiatric ward; even those who had, nevertheless had some genuine fears about what might happen to them. Their nervousness, then, was quite appropriate to the novelty of the hospital setting, and it abated rapidly.

Apart from that short-lived nervousness, the pseudopatient behaved on the ward as he "normally" behaved. The pseudopatient spoke to patients and staff as he might ordinarily. Because there is uncommonly little to do on a psychiatric ward, he attempted to engage others in conversation. When asked by staff how he was feeling, he indicated that he was fine, that he no longer experienced symptoms. He responded to instructions from attendants, to calls for medication (which was not swallowed), and to dining-hall instructions. Beyond such activities as were available to him on the admissions ward, he spent his time writing down his observations about the ward, its patients, and the staff. Initially these notes were written "secretly," but as it soon became clear that no one much cared, they were subsequently written on standard tablets of paper in such public places as the dayroom. No secret was made of these activities.

The pseudopatient, very much as a true psychiatric patient, entered a hospital with no foreknowledge of when he would be discharged. Each was told that he would have to get out by his own devices, essentially by convincing the staff that he was sane. The psychological stresses associated with hospitalization were considerable, and all but one of the pseudopatients desired to be discharged almost immediately after being admitted. They were, therefore, motivated not only to behave sanely, but to be paragons of cooperation. That their behavior was in no way disruptive is confirmed by nursing reports, which have been obtained on most of the patients. These reports uniformly indicate that the patients were "friendly," "cooperative," and "exhibited no abnormal indications."

THE NORMAL ARE NOT DETECTABLY SANE

Despite their public "show" of sanity, the pseudopatients were never detected. Admitted, except in one case, with a diagnosis of schizophrenia, each was discharged with a diagnosis of schizophrenia "in remission." The label "in remission" should in no way be dismissed as a formality, for at no time during any hospitalization had any question been raised about any pseudopatient's simulation. Nor are there any indications in the hospital records that the pseudopatient's status was suspect. Rather, the evidence is strong that, once labeled schizophrenic, the pseudopatient was stuck with that label. If the pseudopatient was to be discharged, he must naturally be "in remission"; but he was not sane, nor, in the institution's view, had he ever been sane.

The uniform failure of recognize sanity cannot be attributed to the quality of the hospitals, for although there were considerable variations among them, several are considered excellent. Nor can it be alleged that there was simply not enough time to observe the pseudopatients. Length of hospitalization ranged from 7 to 52 days, with an average of 19 days. The pseudopatients were not, in fact, carefully observed, but this failure clearly speaks more to traditions within psychiatric hospitals than to lack of opportunity.

Finally, it cannot be said that the failure to recognize the pseudopatients' sanity was due to the fact that they were not behaving sanely. While there was clearly some tension present in all of them, their daily visitors could detect no serious behavioral consequences—nor, indeed, could other patients. It was quite common for the patients to "detect" the pseudopatients' sanity. During the first three hospitalizations, when accurate counts were kept, 35 of a total of 118 patients on the admissions ward voiced their suspicions, some vigorously, "You're not crazy. You're a journalist, or a professor [referring to the continual note-taking]. You're checking up on the hospital." While most of the patients were reassured by the pseudopatient's insistence that he had been sick before he came in but was fine now, some continued to believe that the pseudopatient was sane throughout his hospitalization. The fact that the patients often recognized normality when staff did not raises important questions.

Failure to detect sanity during the course of the hospitalization may be due to the fact that physicians operate with a strong bias toward what statisticians call the type 2 error. This is to say that physicians are more inclined to call a healthy person sick (a false positive, type 2) than a sick person healthy (a false negative, type 1). The reasons for this are not hard to find: it is clearly more dangerous to misdiagnose illness than health. Better to err on the side of caution, to suspect illness even among the healthy.

But what holds for medicine does not hold equally well for psychiatry. Medical illnesses, while unfortunate, are not commonly pejorative. Psychiatric diagnoses, on the contrary, carry with them personal, legal, and social stigmas. It was therefore important to see whether the tendency toward diagnosing the sane insane could be reversed. The following experiment was arranged at a research and teaching hospital whose staff had heard these findings but doubted that such an error could occur in their hospital. The staff was informed that at some time during the following 3 months, one or more pseudopatients would attempt to be admitted into the psychiatric hospital. Each staff member was asked to rate each patient who presented himself at admissions or on the ward according to the likelihood that the patient was a pseudopatient. A 10-point scale was used, with a 1 and 2 reflecting high confidence that the patient was a pseudopatient.

Judgments were obtained on 193 patients who were admitted for psychiatric treatment. All staff who had had sustained contact with or primary responsibility for the patient—attendants, nurses, psychiatrists, physicians, and psychologists—were asked to make judgments. Forty-one patients were alleged, with high confidence, to be pseudopatients by at least one member of the staff. Twenty-three were considered suspect by at least one psychiatrist. Nineteen were suspected by one psychiatrist *and* one other staff member. Actually, no genuine pseudopatient (at least from my group) presented himself during this period.

The experiment is instructive. It indicates that the tendency to designate sane people as insane can be reversed when the stakes (in this case, prestige and diagnostic acumen) are high. But what can be said of the 19 people who were suspected of being "sane" by one psychiatrist and another staff member? Were these people truly "sane," or was it rather the case that in the course of avoiding the type 2 error of the staff tended to make more errors of the first sort—calling the crazy "sane"? There is no way of knowing. But one thing is certain: any diagnostic process that lends itself so readily to massive errors of this sort cannot be a very reliable one.

THE STICKINESS OF PSYCHODIAGNOSTIC LABELS

Beyond the tendency to call the healthy sick—a tendency that accounts better for diagnostic behavior on admission than it does for such behavior after a lengthy period of exposure—the data speak to the massive role of labeling in psychiatric assessment. Having once been labeled schizophrenic, there is nothing the pseudopatient can do to overcome the tag. The tag profoundly colors others' perceptions of him and his behavior....

Once a person is designated abnormal, all of his other behaviors and characteristics are colored by that label. Indeed, that label is so powerful that many of the pseudopatients' normal behaviors were overlooked entirely or profoundly misinterpreted. Some examples may clarify this issue.

Earlier I indicated that there were no changes in the pseudopatient's personal history and current status beyond those of name, employment, and, where necessary, vocation. Otherwise, a veridical description of personal history and circumstances was offered. Those circumstances were not psychotic. How were they made consonant with the diagnosis of psychosis? Or were those diagnoses modified in such a way as to bring them into accord with the circumstances of the pseudopatient's life, as described by him?

As far as I can determine, diagnoses were in no way affected by the relative health of the circumstances of a pseudopatient's life. Rather, the reverse occurred: the perception of his circumstances was shaped entirely by the diagnosis. A clear example of such translation is found in the case of a pseudopatient who had had a close relationship with his mother but was rather remote from his father during his early childhood. During adolescence and

beyond, however, his father became a close friend, while his relationship with his mother cooled. His present relationship with his wife was characteristically close and warm. Apart from occasional angry exchanges, friction was minimal. The children had rarely been spanked. Surely there is nothing especially pathological about such a history. Indeed, many readers may see a similar pattern in their own experiences, with no markedly deleterious consequences. Observe, however, how such a history was translated in the psychopathological context, this from the case summary prepared after the patient was discharged.

> This white 39-year-old male ... manifests a long history of considerable ambivalence in close relationships, which begins in early childhood. A warm relationship with his mother cools during his adolescence. A distant relationship to his father is described as becoming very intense. Affective stability is absent. His attempts to control emotionality with his wife and children are punctuated by angry outbursts and, in the case of the children, spankings. And while he says that he has several good friends, one senses considerable ambivalence embedded in those relationships also....

The facts of the case were unintentionally distorted by the staff to achieve consistency with a popular theory of the dynamics of a schizophrenic reaction. Nothing of an ambivalent nature has been described in relations with parents, spouse, or friends. To the extent that ambivalence could be inferred, it was probably not greater than is found in all human relationships. It is true the pseudopatient's relationships with his parents changed over time, but in the ordinary context that would hardly be remarkable—indeed, it might very well be expected. Clearly, the meaning ascribed to his verbalizations (that is, ambivalence, affective instability) was determined by the diagnosis: schizophrenia. An entirely different meaning would have been ascribed if it were known that the man was "normal."

All pseudopatients took extensive notes publicly. Under ordinary circumstances, such behavior would have raised questions in the minds of observers, as, in fact, it did among patients. Indeed, it seemed so certain that the notes would elicit suspicion that elaborate precautions were taken to remove them from the ward each day. But the precautions proved needless. The closest any staff member came to questioning these notes occurred when one pseudopatient asked his physician what kind of medication he was receiving and began to write down the response. "You needn't write it," he was told gently. "If you have trouble remembering, just ask me again."

If no questions were asked of the pseudopatients, how was their writing interpreted? Nursing records for three patients indicate that the writing was seen as an aspect of their pathological behavior. "Patient engages in writing behavior" was the daily nursing comment on one of the pseudopatients who was never questioned about his writing. Given that the patient is in the hospital, he must be psychologically disturbed. And given that he is disturbed, continuous writing must be a behavioral manifestation of that disturbance, perhaps a subset of the compulsive behaviors that are sometimes correlated with schizophrenia.

One tacit characteristic of the psychiatric diagnosis is that it locates the sources of aberration within the individual and only rarely within the complex of stimuli that surrounds him. Consequently, behaviors that are stimulated by the environment are commonly misattributed to the patient's disorder. For example, one kindly nurse found a pseudopatient pacing the long hospital corridors. "Nervous, Mr. X?" she asked. "No, bored," he said.

The notes kept by pseudopatients are full of patient behaviors that were misinterpreted by well-intentioned staff. Often enough, a patient would go "berserk" because he had, wittingly or unwittingly, been mistreated by, say, an attendant. A nurse coming upon the scene would rarely inquire even cursorily into the environmental stimuli of the patient's behavior. Rather, she assumed that his upset derived from his pathology, not from his present interactions with other staff members. Occasionally, the staff might assume that the patient's family (especially when they had recently visited) or other patients had stimulated the outburst. But never were the staff found to assume that one of themselves or the structure of the hospital had anything to do with a patient's behavior. One psychiatrist pointed to a group of patients who were sitting outside the cafeteria entrance half an hour before lunchtime. To a group of young residents he indicated that such behavior was characteristic of the oral-acquisitive nature of the syndrome. It seemed not to occur to him that there were very few things to anticipate in a psychiatric hospital besides eating.

A psychiatric label has a life and an influence of its own. Once the impression has been formed that the patient is schizophrenic, the expectation is that he will continue to be schizophrenic. When a sufficient amount of time has passed, during which the patient has done nothing bizarre, he is considered to be in remission and available for discharge. But the label endures beyond discharge, with the unconfirmed expectation that he will behave as a schizophrenic again. Such labels, conferred by mental health professionals, are as influential on the patient as they are on his relatives and friends, and it should not surprise anyone that the diagnosis acts on all of them as a self-fulfilling prophecy. Eventually, the patient himself accepts the diagnosis with all of its surplus meanings and expectations, and behaves accordingly.

The inferences to be made from these matters are quite simple. Much as Zigler and Phillips have demonstrated that there is enormous overlap in the symptoms presented by patients who have been variously diagnosed, so there is enormous overlap in the behaviors of the sane and the insane. The sane are not "sane" all of the time. We lose our tempers "for no good reason." We are occasionally depressed or anxious, again for no good reason. And we may find it difficult to get along with one or another person—again for no reason that we can specify. Similarly, the insane are not always insane. Indeed, it was the impression of the pseudopatients while living with them that they were sane for long periods of time—that the bizarre behaviors upon which their diagnoses were allegedly predicated constituted only a small fraction of their total behavior. If it makes no sense to label ourselves permanently depressed on the basis of an occasional depression, then it takes better evidence than is presently available to label all patients insane or schizophrenic on the basis of bizarre behaviors or cognitions. It seems more useful, as Mischel has pointed out, to limit our discussions to *behavior*, the stimuli that provoke them, and their correlates.

It is not known why powerful impressions of personality traits, such as "crazy" or "insane," arise. Conceivably, when the origins of and stimuli that give rise to a behavior are remote or unknown, or when the behavior strikes us as immutable, trait labels regarding the *behaver* arise. When, on the other hand, the origins and stimuli are known and available, discourse is limited to the behavior itself. Thus, I may hallucinate because I am sleeping, or I may hallucinate because I have ingested a peculiar drug. These are termed sleep-induced hallucinations, or dreams, and drug-induced hallucinations, respectively. But when the stimuli to my hallucinations are unknown, that is called craziness, or schizophrenia—as if that inference were somehow as illuminating as the others.

* * *

THE CONSEQUENCES
OF LABELING ...

Whenever the ratio of what is known to what needs to be known approaches zero, we tend to invent "knowledge" and assume that we understand more than we actually do. We seem unable to acknowledge that we simply don't know. The needs for diagnosis and remediation of behavioral and emotional problems are enormous. But rather than acknowledge that we are just embarking on understanding, we continue to label patients "schizophrenic," "manic-depressive," and "insane," as if in those words we had captured the essence of understanding. The facts of the matter are that we have known for a long time that diagnoses are often not useful or reliable, but we have nevertheless continued to use them. We now know that we cannot distinguish insanity from sanity. It is depressing to consider how that information will be used.

Not merely depressing, but frightening. How many people, one wonders, are sane but not recognized as such in our psychiatric institutions? How many have been needlessly stripped of their privileges of citizenship, from the right to vote and drive to that of handling their own accounts. How many have feigned insanity in order to avoid the criminal consequences of their behavior, and, conversely, how many would rather stand trial than live interminably in a psychiatric hospital—but are wrongly thought to be mentally ill? How many have been stigmatized by well-intentioned, but nevertheless erroneous, diagnoses? On the last point, recall again that a "type 2 error" in psychiatric diagnosis does not have the same consequences it does in medical diagnosis. A diagnosis of cancer that has been found to be in error is cause for celebration. But psychiatric diagnoses are rarely found to be in error. The label sticks, a mark of inadequacy forever.

Finally, how many patients might be "sane" outside the psychiatric hospital but seem insane in it—not because craziness resides in them, as it were, but because they are responding to a bizarre setting, one that may be unique to institutions which harbor neither people? Goffman calls the process of socialization to such institutions "mortification."... And while it is impossible to known whether the pseudopatients' responses to these processes are characteristic of all inmates—they were, after all, not real patients—it is difficult to believe that these processes of socialization to a psychiatric hospital provide useful attitudes or habits of response for living in the "real world."

SUMMARY AND
CONCLUSIONS

It is clear that we cannot distinguish the sane from the insane in psychiatric hospitals. The hospital itself imposes a special environment in which the meanings of behavior can easily be misunderstood. The consequences to patients hospitalized in such an environment—the powerlessness, depersonalization, segregation, mortification, and self-labeling—seem undoubtedly counter-therapeutic.

I do not, even now, understand this problem well enough to perceive solutions. But two matters seem to have some promise. The first concerns the proliferation of community mental health facilities, of crisis intervention centers, of the human potential movement, and of behavior therapies that, for all of their own problems, tend to avoid psychiatric labels, to focus on specific problems and behaviors, and to retain the individual in a relatively nonpejorative environment. Clearly, to the extent that we refrain from sending the distressed to insane places, our impressions of them are less likely to be distorted. (The risk of distorted perceptions, it seems to me, is always present, since we are much more sensitive to an individual's behaviors and verbalizations than we are to the subtle contextual stimuli that often promote them. At issue here is a matter of magnitude. And, as I have shown, the magnitude of distortion is exceedingly high in the extreme context that is a psychiatric hospital.)

The second matter that might prove promising speaks to the need to increase the sensitivity of mental health workers and researchers to the *Catch 22* position of psychiatric patients. Simply reading materials in this area will be of help to some such workers and researchers. For others, directly experiencing the impact of psychiatric hospitalization will be of enormous use. Clearly, further research into the social psychology of such total institutions will both facilitate treatment and deepen understanding.

I and the other pseudopatients in the psychiatric setting had distinctly negative reactions. We do not pretend to describe the subjective experiences of true patients. Theirs may be different from ours, particularly with the passage of time and the necessary process of adaptation to one's environment. But we can and do speak to the relatively more objective indices of treatment within the hospital. It could be a mistake, and a very unfortunate one, to consider that what happened to us derived from malice or stupidity on the part of the staff. Quite the contrary, our overwhelming impression of them was of people who really cared, who were committed and who were uncommonly intelligent. Where they failed, as they sometimes did painfully, it would be more accurate to attribute those failures to the environment in which they, too, found themselves than to personal callousness. Their perceptions and behavior were controlled by the situation, rather than being motivated by a malicious disposition. In a more benign environment, one that was less attached to global diagnosis, their behaviors and judgments might have been more benign and effective.

DISCUSSION QUESTIONS

1. D. L. Rosenhan's experiment indicates the importance of labeling the social construction of reality. That is, what is perceived as "real" is the result of people's reactions to certain behaviors. How does this explain other situations in which labeling is an ongoing part of social interaction?

2. In what kinds of environments other than hospitals and asylums might labeling take on the significance that it has in Rosenhan's study? What are the consequences?

3. Goffman's work on the presentation of self (see Reading 7) suggests that people can manipulate how they appear to others, but Rosenhan reminds us that there is always a risk of distorted perceptions by others. How might this explain how people are stereotyped by different social factors?

INTERNET RESOURCES

Suggested Web URLs for Further Study

http://www.stigma.org
This website explores models of psychopathology and their relationship to stigmatization.

http://www.homeoint.org/morrell/otherarticles/marginality.htm
This site contains a presentation given at a conference on sociology on the subject of deviance, marginality, and social exclusion.

http://www.awa.com/norton/struc/chap_18/chap_18.html
This website on psychopathology has links to various forms of mental illness and disorders.

Virtual Society

For further information on this subject, including links to relevant websites, go to the Wadsworth Sociology homepage at *http://www.cengage.com/sociology*.

10

The Uses of Poverty: The Poor Pay All

HERBERT J. GANS

S ome years ago Robert K. Merton applied the notion of functional analysis to explain the continuing though maligned existence of the urban political machine: If it continued to exist, perhaps it fulfilled latent—unintended or unrecognized—positive functions. Clearly it did. Merton pointed out how the political machine provided central authority to get things done when a decentralized local government could not act, humanized the services of the impersonal bureaucracy for fearful citizens, offered concrete help (rather than abstract law or justice) to the poor, and otherwise performed services needed or demanded by many people but considered unconventional or even illegal by formal public agencies.

Today, poverty is more maligned than the political machine ever was; yet it, too, is a persistent social phenomenon. Consequently, there may be some merit in applying functional analysis to poverty, in asking whether it also has positive functions that explain its persistence.

Merton defined functions as "those observed consequences [of a phenomenon] which make for the adaptation of adjustment of a given [social] system." I shall use a slightly different definition; instead of identifying functions for an entire social system, I shall identify them for the interest groups, socioeconomic classes, and other population aggregates with shared values that "inhabit" a social system. I suspect that in a modern heterogeneous society, few phenomena are functional or dysfunctional for the society as a whole, and that most result in benefits to some groups and costs to others. Nor are any phenomena indispensable; in most instances, one can suggest what Merton calls "functional alternatives" or equivalents for them, i.e., other social patterns or policies that achieve the same positive functions but

SOURCE: "The Uses of Poverty," by Herbert Gans from *Social Policy*, July/Aug. 1971, pp. 20–24, © 1971 by Social Policy Corp.

avoid the dysfunction. (In the following discussion, positive functions will be abbreviated as functions and negative functions as dysfunctions. Functions and dysfunctions, in the planner's terminology, will be described as benefits and costs.)

Associating poverty with positive functions seems at first glance to be unimaginable. Of course, the slumlord and the loan shark are commonly known to profit from the existence of poverty, but they are viewed as evil men, so their activities are classified among the dysfunctions of poverty. However, what is less often recognized, at least by the conventional wisdom, is that poverty also makes possible the existence or expansion of respectable professions and occupations, for example, penology, criminology, social work, and public health. More recently, the poor have provided jobs for professional and para-professional "poverty warriors," and for journalists and social scientists, this author included, who have supplied the information demanded by the revival of public interest in poverty.

Clearly, then, poverty and the poor may well satisfy a number of positive functions for many non-poor groups in American society. I shall describe 13 such functions—economic, social, and political—that seem to me most significant.

THE FUNCTIONS OF POVERTY

First, the existence of poverty ensures that society's "dirty work" will be done. Every society has such work: physically dirty or dangerous, temporary, deadend and underpaid, undignified, and menial jobs. Society can fill these jobs by paying higher wages than for "clean" work, or it can force people who have no other choice to do the dirty work—and at low wages. In America, poverty functions to provide a low-wage labor pool that is willing—or, rather, unable to be *un*willing—to perform dirty work at low cost. Indeed, this function of the poor is so important that in some Southern states, welfare payments have been cut off during the summer months when the poor are needed to work in the field. Moreover, much of the debate about the Negative Income Tax and the Family Assistance Plan has concerned their impact on the work incentive, by which is actually meant the incentive of the poor to do the needed dirty work if the wages therefrom are no larger than the income grant. Many economic activities that involve dirty work depend on the poor for their existence: restaurants, hospitals, parts of the garment industry, and "truck farming," among others, could not persist in their present form without the poor.

Second, because the poor are required to work at low wages, they subsidize a variety of economic activities that benefit the affluent. For example, domestics subsidize the upper-middle and upper classes, making life easier for their employers and freeing affluent women for a variety of professional, cultural, civic, and partying activities. Similarly, because the poor pay a higher proportion of their income in property and sales taxes, among others, they subsidize many state and local governmental services that benefit more affluent groups. In addition, the poor support innovation in medical practice as patients in teaching and research hospitals and as guinea pigs in medical experiments.

Third, poverty creates jobs for a number of occupations and professions that serve or "service" the poor, or protect the rest of society from them. As already noted, penology would be minuscule without the poor, as would the police. Other activities and groups that flourish because of the existence of poverty are the numbers game, the sale of heroin and cheap wines and liquors, pentecostal ministers, faith healers, prostitutes, pawn shops, and the peacetime army, which recruits its enlisted men mainly from among the poor.

Fourth, the poor buy goods others do not want and thus prolong the economic usefulness of such goods—day-old bread, fruit and vegetables that would otherwise have to be thrown out, secondhand clothes, and deteriorating automobiles and buildings. They also provide incomes for doctors, lawyers, teachers and others who are too old, poorly trained, or incompetent to attract more affluent clients.

In addition to economic functions, the poor perform a number of social functions.

Fifth, the poor can be identified and punished as alleged or real deviants in order to uphold the legitimacy of conventional norms. To justify the desirability of hard work, thrift, honesty, and monogamy, for example, the defenders of these norms must be able to find people who can be accused of being lazy, spendthrift, dishonest, and promiscuous. Although there is some evidence that the poor are about as moral and law-abiding as anyone else, they are more likely than middle-class transgressors to be caught and punished when they participate in deviant acts. Moreover, they lack the political and cultural power to correct the stereotypes that other people hold of them and thus continue to be thought of as lazy, spendthrift, etc., by those who need living proof that moral deviance does not pay.

Sixth, and conversely, the poor offer vicarious participation to the rest of the population in the uninhibited sexual, alcoholic, and narcotic behavior in which they are alleged to participate and which, being freed from the constraints of affluence, they are often thought to enjoy more than the middle classes. Thus many people, some social scientists included, believe that the poor not only are more given to uninhibited behavior (which may be true, although it is often motivated by despair more than by lack of inhibition) but derive more pleasure from it than affluent people (which research by Lee Rainwater, Walter Miller, and others shows to be patently untrue). However, whether the poor actually have more sex and enjoy it more is irrelevant; so long as middle-class people believe this to be true, they can participate in it vicariously when instances are reported in factual or fictional form.

Seventh, the poor also serve a direct cultural function when culture created by or for them is adopted by the more affluent. The rich often collect artifacts from extinct folk cultures of poor people; and almost all Americans listen to the blues, Negro spirituals, and country music, which originated among the Southern poor. Recently they have enjoyed the rock styles that were born, like the Beatles, in the slums; and in the last year, poetry written by ghetto children has become popular in literary circles. The poor also serve as culture heroes, particularly, of course, to the left; but the hobo, the cowboy, the hipster, and the mythical prostitute

with a heart of gold have performed this function for a variety of groups.

Eighth, poverty helps to guarantee the status of those who are not poor. In every hierarchical society someone has to be at the bottom; but in American society, in which social mobility is an important goal for many and people need to know where they stand, the poor function as a reliable and relatively permanent measuring rod for status comparisons. This is particularly true for the working class, whose politics is influenced by the need to maintain status distinctions between themselves and the poor, much as the aristocracy must find ways of distinguishing itself from the *nouveaux riches.*

Ninth, the poor also aid the upward mobility of groups just above them in the class hierarchy. Thus a goodly number of Americans have entered the middle class through the profits earned from the provision of goods and services in the slums, including illegal or nonrespectable ones that upper-class and upper-middle-class businessmen shun because of their low prestige. As a result, members of almost every immigrant group have financed their upward mobility by providing slum housing, entertainment, gambling, narcotics, etc., to later arrivals—most recently to blacks and Puerto Ricans.

Tenth, the poor help to keep the aristocracy busy, thus justifying its continued existence. "Society" uses the poor as clients of settlement houses and beneficiaries of charity affairs; indeed, the aristocracy must have the poor to demonstrate its superiority over other elites who devote themselves to earning money.

Eleventh, the poor, being powerless, can be made to absorb the costs of change and growth in American society. During the nineteenth century, they did the backbreaking work that built the cities; today, they are pushed out of their neighborhoods to make room for "progress." Urban renewal projects to hold middle-class taxpayers in the city and expressways to enable suburbanites to commute downtown have typically been located in poor neighborhoods, since no other group will allow itself to be displaced. For the same reason, universities, hospitals, and civic centers also expand into land occupied by the poor. The major costs of the industrialization of agriculture have been borne by the poor, who are pushed off the land without recompense; and they have paid a large share of the human cost of the growth of American power overseas, for they have provided many of the foot soldiers for Vietnam and other wars.

Twelfth, the poor facilitate and stabilize the American political process. Because they vote and participate in politics less than other groups, the political system is often free to ignore them. Moreover, since they can rarely support Republicans, they often provide the Democrats with a captive constituency that has no other place to go. As a result, the Democrats can count on their votes, and be more responsive to voters—for example, the white working class—who might otherwise switch to the Republicans.

Thirteenth, the role of the poor in upholding conventional norms (see the *fifth* point, above) also has a significant political function. An economy based on the ideology of laissez-faire requires a deprived population that is allegedly unwilling to work or that can be considered inferior because it must accept charity or welfare in order to survive. Not only does the alleged moral deviancy of the poor reduce the moral pressure on the present political economy to eliminate poverty, but socialist alternatives can be made to look quite unattractive if those who will benefit most from them can be described as lazy, spendthrift, dishonest, and promiscuous.

THE ALTERNATIVES

I have described 13 of the more important functions poverty and the poor satisfy in American society, enough to support the functionalist thesis that poverty, like any other social phenomenon, survives in part because it is useful to society or some of its parts. This analysis is not intended to suggest that because it is often functional, poverty *should* exist, or that it *must* exist. For one thing, poverty has many more dysfunctions than functions; for another, it is possible to suggest functional alternatives.

For example, society's dirty work could be done without poverty, either by automation or by paying "dirty workers" decent wages. Nor is it necessary for the poor to subsidize the many activities they support through their low-wage jobs. This would, however, drive up the costs of these activities, which would result in higher prices to their customers and clients. Similarly, many of the professionals who flourish because of the poor could be given other roles. Social workers could provide counseling to the affluent, as they prefer to do anyway; and the police could devote themselves to traffic and organized crime. Other roles would have to be found for badly trained or incompetent professionals now relegated to serving the poor, and someone else would have to pay their salaries. Fewer penologists would be employable, however. And pentecostal religion could probably not survive without the poor—nor would parts of the second- and third-hand-goods market. And in many cities, "used" housing that no one else wants would then have to be torn down at public expense.

Alternatives for the cultural functions of the poor could be found more easily and cheaply. Indeed, entertainers, hippies, and adolescents are already serving as the deviants needed to uphold traditional morality and as devotees of orgies to "staff" the fantasies of vicarious participation.

The status functions of the poor are another matter. In a hierarchical society, some people must be defined as inferior to everyone else with respect to a variety of attributes, but they need not be poor in the absolute sense. One could conceive of a society in which the "lower class," though last in the pecking order, received 75 percent of the median income, rather than 15–40 percent, as is now the case. Needless to say, this would require considerable income redistribution.

The contribution the poor make to the upward mobility of the groups that provide them with goods and services could also be maintained without the poor's having such low incomes. However, it is true that if the poor were more affluent, they would have access to enough capital to take over the provider role, thus competing with, and perhaps rejecting, the "outsiders." (Indeed, owing in part to antipoverty programs, this is already happening in a number of ghettos, where white storeowners are being replaced by blacks.) Similarly, if the poor were more affluent, they would make less willing clients for upper-class philanthropy, although some would still use settlement houses to achieve upward mobility, as they do now. Thus "Society" could continue to run its philanthropic activities.

The political functions of the poor would be more difficult to replace. With increased affluence the poor would probably obtain more political power and be more active politically. With higher incomes and more political power, the poor would be likely to resist paying the costs of growth and change. Of course, it is possible to imagine urban renewal and highway projects that properly reimbursed the displaced people, but such projects would then become considerably more expensive, and many might never be built. This, in turn, would reduce the comfort and convenience of those who now benefit from urban renewal and express ways. Finally, hippies could serve also as more deviants to justify the existing political economy—as they already do. Presumably, however, if poverty were eliminated, there would be fewer attacks on that economy.

In sum, then, many of the functions served by the poor could be replaced if poverty were eliminated, but almost always at higher costs to others, particularly more affluent others. Consequently, a functional analysis must conclude that poverty persists not only because it fulfills a number of positive functions but also because many of the functional alternatives to poverty would be quite dysfunctional for the affluent members of society. A functional analysis thus ultimately arrives at much the same conclusion as radical sociology, except that radical thinkers treat as manifest what I describe as latent: that social phenomena that are functional for affluent or powerful groups and dysfunctional for poor or powerless ones persist; that when the elimination of such phenomena through functional alternatives would generate dysfunctions for the affluent or powerful, they will continue to persist; and that phenomena like poverty can be eliminated only when they become dysfunctional for the affluent or powerful, or when the powerless can obtain enough power to change society.

POSTSCRIPT

Over the years, this article has been interpreted as either a direct attack on functionalism or a tongue-in-cheek satirical comment on it. Neither interpretation is true. I wrote the article for two reasons. First and foremost, I wanted to point out that there are, unfortunately, positive functions of poverty which have to be dealt with by antipoverty policy. Second, I was trying to show that functionalism is not the inherently conservative approach for which it has often been criticized, but that it can be employed in liberal and radical analyses.

DISCUSSION QUESTIONS

1. What specific groups does Herbert J. Gans identify as benefiting from poverty?

2. Gans wrote this essay in 1971. Had he written it today, what specific observations about the functions of poverty might he further identify? Be as specific as possible.

3. Gans uses the perspective of functionalist theory to think about poverty; at the same time, he draws from conflict theory. What do each of these perspectives contribute to an understanding of poverty in society?

INTERNET RESOURCES

Suggested Web URLs for Further Study

http://www.ssc.wisc.edu/irp/
This is the official site for the Institute for Research on Poverty. IRP is a national nonprofit and nonpartisan organization dedicated to researching the causes and consequences of poverty and social inequality in the United States.

Virtual Society

For further information on this subject, including links to relevant websites, go to the Wadsworth Sociology homepage at *http://www.cengage.com/sociology*.

11

The Souls of Black Folk

W. E. B. DU BOIS

Between me and the other world there is ever an unasked question; unasked by some through feelings of delicacy; by others through the difficulty of rightly framing it. All, nevertheless, flutter round it. They approach me in a half-hesitant sort of way, eye me curiously or compassionately, and then, instead of saying directly, How does it feel to be a problem? they say, I know an excellent colored man in my town; or, I fought at Mechanicsville; or, Do not these Southern outrages make your blood boil? At these I smile, or am interested, or reduce the boiling to a simmer, as the occasion may require. To the real question, How does it feel to be a problem? I answer seldom a word.

And yet, being a problem is a strange experience,—peculiar even for one who has never been anything else, save perhaps in babyhood and in Europe. It is in the early days of rollicking boyhood that the revelation first bursts upon one, all in a day, as it were. I remember well when the shadow swept across me. I was a little thing, away up in the hills of New England, where the dark Housatonic winds between Hoosac and Taghkanic to the sea. In a wee wooden schoolhouse, something put it into the boys' and girls' heads to buy gorgeous visiting cards—ten cents a package—and exchange. The exchange was merry, till one girl, a tall newcomer, refused my card—refused it per-emptorily, with a glance. Then it dawned upon me with a certain suddenness that I was different from the others; or like, mayhap, in heart and life and longing, but shut out from their world by a vast veil. I had thereafter no desire to tear down that veil, to creep through; I held all beyond it in common contempt, and lived above it in a region of blue sky and great wandering shadows. That sky was bluest when I could beat my mates at examination-time, or beat them at a foot-race, or even beat their stringy heads. Alas, with the years all this fine contempt began to fade; for the words I longed for, and all their dazzling opportunities, were theirs, not mine. But they should not keep these prizes, I said; some, all, I would wrest from them. Just how I would do it I could never decide: by reading law, by healing the sick, by telling the wonderful tales that swam in my head,—some way. With other black boys the strife was not so fiercely sunny:

Their youth shrunk into tasteless sycophancy, or into silent hatred of the pale world about them and mocking distrust of everything white; or wasted itself in a bitter cry, Why did God make me an outcast and stranger in mine own house? The shades of the prison-house closed round about us all: walls strait and stubborn to the whitest, but relentlessly narrow, tall, and unscalable to sons of night who must plod darkly on in resignation, or beat unavailing palms against the stone, or steadily, half hopelessly, watch the streak of blue above.

After the Egyptian and Indian, the Greek and Roman, the Teuton and Mongolian, the Negro is a sort of seventh son, born with a veil, and gifted with second-sight in this American world—a world which yields him no true self-consciousness, but only lets him see himself through the revelation of the other world. It is a peculiar sensation, this double-consciousness, this sense of always looking at one's self through the eyes of others, of measuring one's soul by the tape of a world that looks on in amused contempt and pity. One ever feels his twoness—an American, a Negro; two souls, two thoughts, two unreconciled strivings; two warring ideals in one dark body, whose dogged strength alone keeps it from being torn asunder.

The history of the American Negro is the history of this strife—this longing to attain self-conscious manhood, to merge his double self into a better and truer self. In this merging he wishes neither of the older selves to be lost. He would not Africanize America, for America has too much to teach the world and Africa. He would not bleach his Negro soul in a flood of white Americanism, for he knows that Negro blood has a message for the world. He simply wishes to make it possible for a man to be both a Negro and an American, without being cursed and spit upon by his fellows, without having the doors of Opportunity closed roughly in his face.

This, then, is the end of his striving: to be a coworker in the kingdom of culture, to escape both death and isolation, to husband and use his best powers and his latent genius. These powers of body and mind have in the past been strangely wasted, dispersed, or forgotten. The shadow of a mighty Negro past flits through the tale of Ethiopia the Shadowy and of Egypt the Sphinx. Through history, the powers of single black men flash here and there like falling stars, and die sometimes before the world has rightly gauged their brightness. Here in America, in the few days since Emancipation, the black man's turning hither and thither in hesitant and doubtful striving has often made his very strength to lose effectiveness, to seem like absence of power, like weakness. And yet it is not weakness—it is the contradiction of double aims. The double-aimed struggle of the black artisan on the one hand to escape white contempt for a nation of mere hewers of wood and drawers of water, and on the other hand to plough and nail and dig for a poverty-stricken horde—could only result in making him a poor craftsman, for he had but half a heart in either cause. By the poverty and ignorance of his people, the Negro minister or doctor was tempted toward quackery and demagogy; and by the criticism of the other world, toward ideals that made him ashamed of his lowly tasks. The would-be black savant was confronted by the paradox that the knowledge his people needed was a twice-told tale to his white neighbors, while the knowledge which would teach the white world was Greek to his own flesh and blood. The innate love of harmony and beauty that set the ruder souls of his people a-dancing and a-singing raised but confusion and doubt in the soul of the black artist; for the beauty revealed to him was the soul-beauty of a race which his larger audience despised, and he could not articulate the message of another people. This waste of double aims, this seeking to satisfy two unreconciled ideals, has wrought sad havoc with the courage and faith and deeds of ten thousand people—has sent them often wooing false gods and invoking false means of salvation, and at times has even seemed about to make them ashamed of themselves.

Away back in the days of bondage they thought to see in one divine event the end of all doubt and disappointment; few men ever worshipped Freedom with half such unquestioning faith as did the American Negro for two centuries. To him, so far as he thought and dreamed, slavery was indeed the sum of all villainies, the cause of all sorrow, the root of all prejudice; Emancipation was the key to a promised land of sweeter beauty than ever stretched before the eyes of wearied Israelites. In song and exhortation swelled one refrain— Liberty; in his tears and curses the God he implored had Freedom in his right hand. At last it came— suddenly, fearfully, like a dream. With one wild carnival of blood and passion came the message in his own plaintive cadences:

> Shout, O children!
> Shout, you're free!
> For God has bought your liberty!

Years have passed away since then—ten, twenty, forty; forty years of national life, forty years of renewal and development, and yet the swarthy spectre sits in its accustomed seat at the Nation's feast. In vain do we cry to this our vastest social problem:

> Take any shape but that, and my firm
> nerves shall never tremble!

The Nation has not yet found peace from its sins; the freedman has not yet found in freedom his promised land. Whatever of good may have come in these years of change, the shadow of a deep disappointment rests upon the Negro people—a disappointment all the more bitter because the unattained ideal was unbounded save by the simple ignorance of a lowly people.

The first decade was merely a prolongation of the vain search for freedom, the boon that seemed ever barely to elude their grasp—like a tantalizing will-o'-the-wisp, maddening and misleading the headless host. The holocaust of war, the terrors of the Ku-Klux Klan, the lies of carpet-baggers, the disorganization of industry, and the contradictory advice of friends and foes, left the bewildered serf with no new watchword beyond the old cry for freedom. As the time flew, however, he began to grasp a new idea. The ideal of liberty demanded for its attainment powerful means, and these the Fifteenth Amendment gave him. The ballot, which before he had looked upon as a visible sign of freedom, he now regarded as the chief means of gaining and perfecting the liberty with which war had partially endowed him. And why not? Had not votes made war and emancipated millions? Had not votes enfranchised the freedmen? Was

anything impossible to a power that had done all this? A million black men started with renewed zeal to vote themselves into the kingdom. So the decade flew away, the revolution of 1876 came, and left the half-free serf weary, wondering, but still inspired. Slowly but steadily, in the following years, a new vision began gradually to replace the dream of political power—a powerful movement, the rise of another ideal to guide the unguided, another pillar of fire by night after a clouded day. It was the ideal of "book-learning"; the curiosity, born of compulsory ignorance, to know and test the power of the cabalistic letters of the white man, the longing to know. Here at last seemed to have been discovered the mountain path to Canaan; longer than the highway of Emancipation and law, steep and rugged, but straight, leading to heights high enough to overlook life.

Up the new path the advance guard toiled, slowly, heavily, doggedly; only those who have watched and guided the faltering feet, the misty minds, the dull understandings, of the dark pupils of these schools know how faithfully, how piteously, this people strove to learn. It was weary work. The cold statistician wrote down the inches of progress here and there, noted also where here and there a foot had slipped or some one had fallen. To the tired climbers, the horizon was ever dark, the mists were often cold, the Canaan was always dim and far away. If, however, the vistas disclosed as yet no goal, no resting-place, little but flattery and criticism, the journey at least gave leisure for reflection and self-examination; it changed the child of Emancipation to the youth with dawning self-consciousness, self-realization, self-respect. In those sombre forests of his striving his own soul rose before him, and he saw himself darkly as through a veil; and yet he saw in himself some faint revelation of his power, of his mission. He began to have a dim feeling that, to attain his place in the world, he must be himself, and not another. For the first time he sought to analyze the burden he bore upon his back, that dead-weight of social degradation partially masked behind a half-named Negro problem. He felt his poverty; without a cent, without a home, without land, tools, or savings, he had entered into competition with rich, landed, skilled neighbors. To be a poor man is hard, but to be a poor race in a land of dollars is the very bottom of hardships. He felt the weight of his ignorance—not simply of letters, but of life, of business, of the humanities; the accumulated sloth and shirking and awkwardness of decades and centuries shackled his hands and feet. Nor was his burgeon all poverty and ignorance. The red stain of bastardy, which two centuries of systematic legal defilement of Negro women had stamped upon his race, meant not only the loss of ancient African chastity, but also the hereditary weight of a mass of corruption from white adulterers, threatening almost the obliteration of the Negro home.

A people thus handicapped ought not to be asked to race with the world, but rather allowed to give all its time and thought to its own social problems. But alas! While sociologists gleefully count his bastards and his prostitutes, the very soul of the toiling, sweating black man is darkened by the shadow of a vast despair. Men call the shadow prejudice, and learnedly explain it as the natural defence of culture against barbarism, learning against ignorance, purity against crime, the "higher" against the "lower" races. To which the Negro cries Amen! and swears that to so much of this strange prejudice as is founded on just homage to civilization, culture, righteousness, and progress, he humbly bows and meekly does obeisance. But before that nameless prejudice that leaps beyond all this he stands helpless, dismayed, and well-nigh speechless; before that personal disrespect and mockery, the ridicule and systematic humiliation, the distortion of fact and wanton license of fancy, the cynical ignoring of the better and the boisterous welcoming of the worse, the all-pervading desire to inculcate disdain for everything black, from Toussaint to the devil—before this there rises a sickening despair that would disarm and discourage any nation save that black host to whom "discouragement" is an unwritten word.

But the facing of so vast a prejudice could not but bring the inevitable self-questioning, self-disparagement, and lowering of ideals which ever accompany repression and breed in an atmosphere of contempt and hate. Whisperings and portents came borne upon the four winds: Lo! we are diseased and dying, cried the dark hosts; we cannot write, our

voting is vain; what need of education, since we must always cook and serve? And the Nation echoed and enforced this self-criticism saying: Be content to be servants, and nothing more; what need of higher culture for half-men? Away with the black man's ballot, by force or fraud—and behold the suicide of a race! Nevertheless, out of the evil came something of good—the more careful adjustment of education to real life, the clear perception of the Negroes' social responsibilities, and the sobering realization of the meaning of progress.

So dawned the time of *Sturm und Drang*: Storm and stress today rocks our little boat on the mad waters of the world-sea; there is within and without the sound of conflict, the burning of body and rending of soul; inspiration strives with doubt, and faith with vain questionings. The bright ideals of the past—physical freedom, political power, the training of brains and the training of hands—all these in turn have waxed and waned, until even the last grows dim and overcast. Are they all wrong—all false? No, not that, but each alone was oversimple and incomplete—the dreams of a credulous race—childhood, or the fond imaginings of the other world which does not know and does not want to know our power. To be really true, all these ideals must be melted and welded into one. The training of the schools we need today more than ever—the training of deft hands, quick eyes and ears, and above all the broader, deeper, higher culture of gifted minds and pure hearts. The power of the ballot we need in sheer self-defense—else what shall save us from a second slavery? Freedom, too, the long-sought, we still seek—the freedom of life and climb, the freedom to work and think, the freedom to love and aspire. Work, culture, liberty—all these we need, not singly but together, not successively but together, each growing and aiding each, and all striving toward that vaster ideal that swims before the Negro people, the ideal of human brotherhood, gained through the unifying ideal of Race; the ideal of fostering and developing the traits and talents of the Negro, not in opposition to or contempt for other races, but rather in large conformity to the greater ideals of the American Republic, in order that some day on American soil two world-races may give each to each those characteristics both so sadly lack. We the darker ones come even now not altogether empty-handed: There are today no truer exponents of the pure human spirit of the Declaration of Independence than the American Negroes; there is no true American music but the wild sweet melodies of the Negro slave, the American fairy tales and folklore are Indian and African; and, all in all, we black men seem the sole oasis of simple faith and reverence in a dusty desert of dollars and smartness. Will America be poorer if she replace her brutal dyspeptic blundering with lighthearted but determined Negro humility? Or her coarse and cruel wit with loving jovial good-humor? Or her vulgar music with the soul of the Sorrow Songs?

Merely a concrete test of the underlying principles of the great republic is the Negro Problem, and the spiritual striving of the freedmen's sons is the travail of souls whose burden is almost beyond the measure of their strength, but who bear it in the name of an historic race, in the name of this the land of their fathers' fathers, and in the name of human opportunity.

DISCUSSION QUESTIONS

1. What are some of the consequences of racial prejudice, according to W. E. B. Du Bois? What evidence do you see of this in contemporary race relations?

2. Du Bois was writing at the beginning of the 20th century. Suppose he returned to the United States now. Following from his original ideas, what do you think he would write now?

3. Du Bois's concept of "the veil" has been used to refer to the double consciousness that African Americans have to have as an oppressed group in society—that is, knowing themselves and knowing the oppressor. Why does this emerge, and what implications does it have for understanding how people's knowledge develops in a racially stratified society?

INTERNET RESOURCES

Suggested Web URLs for Further Study

http://www.lucidcafe.com/library/96feb/dubois.html
This site provides a biography of Du Bois, a list of his books and links to other civil rights leaders.

http://race.eserver.org/resources.html
This site is dedicated to race and ethnicity with a list of resources, articles, essays, texts, narratives, and more.

Virtual Society

For further information on this subject, including links to relevant websites, go to the Wadsworth Sociology homepage at *http://www.cengage.com/sociology*.

12

Women and Economics

CHARLOTTE PERKINS GILMAN

Since we have learned to study the development of human life as we study the evolution of species throughout the animal kingdom, some peculiar phenomena ... begin to show themselves in a new light. We begin to see that, so far from being inscrutable problems, ... these sorrows and perplexities of our lives are but the natural results of natural causes, and that, as soon as we ascertain the causes, we can do much to remove them.

In spite of the power of the individual will to struggle against conditions, ... it remains true that the human creature is affected by his environment, as is every other living thing....

The general course of life shows the inexorable effect of conditions upon humanity. Of these conditions we share with other living things the environment of the material universe. We are affected by climate and locality, by physical, chemical, electrical forces, as are all animals and plants. With the animals, we farther share the effect of our own activity, the reactionary force of exercise. What we do, as well as what is done to us, makes us what we are. But, beyond these forces, we come under the effect of a third set of conditions peculiar to our human status; namely, social conditions. In the organic interchanges which constitute social life, we are affected by each other to a degree beyond what is found even among the most gregarious of animals. This third factor, the social environment, is of enormous force as a modifier of human life. Throughout all these environing conditions, those which affect us through our economic necessities are most marked in their influence.

Without touching yet upon the influence of the social factors, treating the human being merely as an individual animal, we see that he is modified most by his economic conditions, as is every other animal. Differ as they may in color and size, in strength and speed, in minor adaptation to minor conditions, all animals that live on grass have distinctive traits in common, and all animals that eat flesh have distinctive traits in common,—so distinctive and so common that it is by teeth, by nutritive

SOURCE: Charlotte Perkins Gilman. 1898/1998. *Women and Economics*. New York: Dover. Used by permission.

apparatus in general, that they are classified, rather than by means of defence or locomotion. The food supply of the animal is the largest passive factor in his development; the processes by which he obtains his food supply, the largest active factor in his development. It is these activities, the incessant repetition of his exertions by which he is fed, which most modify his structure and develop his functions. The sheep, the cow, the deer, differ in their adaptation to the weather, their locomotive ability, their means of defence; but they agree in main characteristics, because of their common method of nutrition.

The human animal is no exception to this rule. Climate affects him, weather affects him, enemies affect him; but most of all he is affected, like every other living creature, by what he does for his living. Under all the influence of his later and wider life, all the reactive effect of social institutions, the individual is still inexorably modified by his means of livelihood: "the hand of the dyer is subdued to what he works in."...

In view of these facts, attention is now called to a certain marked and peculiar economic condition affecting the human race, and unparalleled in the organic world. We are the only animal species in which the female depends on the male for food, the only animal species in which the sex-relation is also an economic relation. With us an entire sex lives in a relation of economic dependence upon the other sex, and the economic relation is combined with the sex-relation. The economic status of the human female is relative to the sex-relation.

It is commonly assumed that this condition also obtains among other animals, but such is not the case. There are many birds among which, during the nesting season, the male helps the female feed the young, and partially feeds her; and, with certain of the higher carnivora, the male helps the female feed the young, and partially feeds her. In no case does she depend on him absolutely, even during this season, save in that of the hornbill, where the female, sitting on her nest in a hollow tree, is walled in with clay by the male, so that only her beak projects; and then he feeds her while the eggs are developing. But even the female hornbill does not expect to be fed at any other time. The female bee

and ant are economically dependent, but not on the male. The workers are females, too, specialized to economic functions solely. And with the carnivora, if the young are to lose one parent, it might far better be the father: the mother is quite competent to take care of them herself. With many species, as in the case of the common cat, she not only feeds herself and her young, but has to defend the young against the male as well. In no case is the female throughout her life supported by the male.

In the human species the condition is permanent and general, though there are exceptions, and though the present century is witnessing the beginnings of a great change in this respect. We have not been accustomed to face this fact beyond our loose generalization that it was "natural," and that other animals did so, too.

To many this view will not seem clear at first; and the case of working peasant women or females or savage tribes, and the general household industry of women, will be instanced against it. Some careful and honest discrimination is needed to make plain to ourselves the essential facts of the relation, even in these cases. The horse, in his free natural condition, is economically independent. He gets his living by his own exertions, irrespective of any other creature. The horse, in his present condition of slavery, is economically dependent. He gets his living at the hands of his master; and his exertions, though strenuous, bear no direct relation to his living. In fact, the horses who are the best fed and cared for and the horses who are the hardest worked are quite different animals. The horse works, it is true; but what he gets to eat depends on the power and will of his master. His living comes through another. He is economically dependent. So with the hard-worked savage or peasant women. Their labor is the property of another: they work under another will; and what they receive depends not on their labor, but on the power and will of another. They are economically dependent. This is true of the human female both individually and collectively.

In studying the economic position of the sexes collectively, the difference is most marked. As a social animal, the economic status of man rests on the combined and exchanged services of vast numbers of progressively specialized individuals. The economic progress of the race, its maintenance at any period, its continued advance, involve the collective activities of all the trades, crafts, arts, manufactures, inventions, discoveries, and all the civil and military institutions that go to maintain them. The economic status of any race at any time, with its involved effect on all the constituent individuals, depends on their world-wide labors and their free exchange. Economic progress, however, is almost exclusively masculine. Such economic processes as women have been allowed to exercise are of the earliest and most primitive kind. Were men to perform no economic services save such as are still performed by women, our racial status in economics would be reduced to most painful limitations.

To take from any community its male workers would paralyze it economically to a far greater degree than to remove its female workers. The labor now performed by the women could be performed by the men, requiring only the setting back of many advanced workers into earlier forms of industry; but the labor now performed by the men could not be performed by the women without generations of effort and adaptation. Men can cook, clean, and sew as well as women; but the making and managing of the great engines of modern industry, the threading of earth and sea in our vast systems of transportation, the handling of our elaborate machinery of trade, commerce, government,—these things could not be done so well by women in their present degree of economic development.

This is not owing to lack of the essential human faculties necessary to such achievements, nor to any inherent disability of sex, but to the present condition of woman, forbidding the development of this degree of economic ability. The male human being is thousands of years in advance of the female in economic status. Speaking collectively, men produce and distribute wealth; and women receive it at their hands. As men hunt, fish, keep cattle, or raise corn, so do women eat game, fish, beef, or corn. As men go down to the sea in ships, and bring coffee and spices and silks and gems from far away, so do women partake of the coffee and spices and silks and gems the men bring.

The economic status of the human race in any nation, at any time, is governed mainly by the activities of the male: the female obtains her share in the racial advance only through him.

Studied individually, the facts are even more plainly visible, more open and familiar. From the day laborer to the millionaire, the wife's worn dress or flashing jewels, her low roof or her lordly one, her weary feet or her rich equipage,—these speak of the economic ability of the husband. The comfort, the luxury, the necessities of life itself, which the woman receives, are obtained by the husband, and given her by him. And, when the woman, left alone with no man to "support" her, tries to meet her own economic necessities, the difficulties which confront her prove conclusively what the general economic status of the woman is. None can deny these patent facts,— that the economic status of women generally depends upon that of men generally, and that the economic status of women individually depends upon that of men individually, those men to whom they are related. But we are instantly confronted by the commonly received opinion that, although it must be admitted that men make and distribute the wealth of the world, yet women earn their share of it as wives. This assumes either that the husband is in the position of employer and the wife as employee, or that marriage is a "partnership," and the wife an equal factor with the husband in producing wealth.

Economic independence is a relative condition at best. In the broadest sense, all living things are economically dependent upon others,—the animals upon the vegetables, and man upon both. In a narrower sense, all social life is economically interdependent, man producing collectively what he could by no possibility produce separately. But, in the closest interpretation, individual economic independence among human beings means that the individual pays for what he gets, works for what he gets, gives to the other an equivalent for what the other gives him. I depend on the shoemaker for shoes, and the tailor for coats; but, if I give the shoemaker and the tailor enough of my own labor as a house-builder to pay for the shoes and coats they give me, I retain my personal independence. I have not taken of their product, and given nothing of mine. As long as what I get is obtained by what I give. I am economically independent.

Women consume economic goods. What economic product do they give in exchange for what they consume? The claim that marriage is a partnership, in which the two persons married produce wealth which neither of them, separately, could produce, will not bear examination. A man happy and comfortable can produce more than one unhappy and uncomfortable, but this is as true of a father or son as of a husband. To take from a man any of the conditions which make him happy and strong is to cripple his industry, generally speaking. But those relatives who make him happy are not therefore his business partners, and entitled to share his income.

Grateful return for happiness conferred is not the method of exchange in a partnership. The comfort a man takes with his wife is not in the nature of a business partnership, nor are her frugality and industry. A housekeeper, in her place, might be as frugal, as industrious, but would not therefore be a partner. Man and wife are partners truly in their mutual obligation to their children,—their common love, duty, and service. But a manufacturer who marries, or a doctor, or a lawyer, does not take a partner in his business, when he takes a partner in parenthood, unless his wife is also a manufacturer, a doctor, or a lawyer. In his business, she cannot even advise wisely without training and experience. To love her husband, the composer, does not enable her to compose; and the loss of a man's wife, though it may break his heart, does not cripple his business, unless his mind is affected by grief. She is in no sense a business partner, unless she contributes capital or experience or labor, as a man would in like relation. Most men would hesitate very seriously before entering a business partnership with any woman, wife or not.

If the wife is not, then, truly a business partner, in what way does she earn from her husband the food, clothing, and shelter she receives at his hands? By house service, it will be instantly replied. This is the general misty idea upon the subject,—that women earn all they get, and more, by house service. Here we come to a very practical and definite economic ground. Although not producers of wealth, women serve in the final processes of preparation and distribution. Their labor in the household has a genuine economic value.

For a certain percentage of persons to serve other persons, in order that the ones so served may produce more, is a contribution not to be overlooked. The labor of women in the house, certainly, enables men to produce more wealth than they otherwise could; and in this way women are economic factors in society. But so are horses. The labor of horses enables men to produce more wealth than they otherwise could. The horse is an economic factor in society. But the horse is not economically independent, nor is the women. If a man plus a valet can perform more useful service than he could minus a valet, then the valet is performing useful service. But, if the valet is the property of the man, is obliged to perform this service, and is not paid for it, he is not economically independent.

The labor which the wife performs in the household is given as part of her functional duty, not as employment. The wife of the poor man, who works hard in a small house, doing all the work for the family, or the wife of the rich man, who wisely and gracefully manages a large house and administers its functions, each is entitled to fair pay for services rendered.

To take this ground and hold it honestly, wives, as earners through domestic service, are entitled to the wages of cooks, housemaids, nurse-maids, seamstresses, or housekeepers, and to no more. This would of course reduce the spending money of the wives of the rich, and put it out of the power of the poor man to "support" a wife at all, unless, indeed, the poor man faced the situation fully, paid his wife her wages as house servant, and then she and he combined their funds in the support of their children. He would be keeping a servant: she would be helping keep the family. But nowhere on earth would there be "a rich woman" by these means. Even the highest class of private housekeeper, useful as her services are, does not accumulate a fortune. She does not buy diamonds and sables and keep a carriage. Things like these are not earned by house service.

But the salient fact in this discussion is that, whatever the economic value of the domestic industry of women is, they do not get it. The women who do the most work get the least money, and the women who have the most money do the least work. Their labor is neither given nor taken as a factor in economic exchange. It is held to be their duty as women to do this work; and their economic status bears no relation to their domestic labors, unless an inverse one. Moreover, if they were thus fairly paid,—given what they earned, and no more,—all women working in this way would be reduced to the economic status of the house servant. Few women—or men either—care to face this condition. The ground that women earn their living by domestic labor is instantly forsaken, and we are told that they obtain their livelihood as mothers. This is a peculiar position. We speak of it commonly enough, and often with deep feeling, but without due analysis.

In treating of an economic exchange, asking what return in goods or labor women make for the foods and labor given them,—either to the race collectively or to their husbands individually,—what payment women make for their clothes and shoes and furniture and food and shelter, we are told that the duties and services of the mother entitle her to support.

If this is so, if motherhood is an exchangeable commodity given by women in payment for clothes and food, then we must of course find some relation between the quantity or quality of the motherhood and the quantity and quality of the pay. This being true, then the women who are not mothers have no economic status at all; and the economic status of those who are must be shown to be relative to their motherhood. This is obviously absurd. The childless wife has as much money as the mother of many,—more; for the children of the latter consume what would otherwise be hers; and the inefficient mother is no less provided for than the efficient one. Visibly, and upon the face of it, women are not maintained in economic prosperity proportioned to their motherhood. Motherhood bears no relation to their economic status. Among primitive races, it is true,—in the patriarchal period, for instance,—there was some truth in this position. Women being of no value whatever save as bearers of children, their favor and indulgence did bear direct relation to maternity;

and they had reason to exult on more grounds than one when they could boast a son. Today, however, the maintenance of the woman is not conditioned upon this. A man is not allowed to discard his wife because she is barren. The claim of motherhood as a factor in economic exchange is false today. But suppose it were true. Are we willing to hold this ground, even in theory? Are we willing to consider motherhood as a business, a form of commercial exchange? Are the cares and duties of the mother, her travail and her love, commodities to be exchanged for bread?

It is revolting so to consider them; and, if we dare face our own thoughts, and force them to their logical conclusion, we shall see that nothing could be more repugnant to human feeling, or more socially and individually injurious, than to make motherhood a trade. Driven off these alleged grounds of women's economic independence; shown that women, as a class, neither produce nor distribute wealth; that women, as individuals, labor mainly as house servants, are not paid as such, and would not be satisfied with such an economic status if they were so paid; that wives are not business partners or co-producers of wealth with their husbands, unless they actually practise the same profession; that they are not salaried as mothers, and that it would be unspeakably degrading if they were,—what remains to those who deny that women are supported by men? This (and a most amusing position it is),—that the function of maternity unfits a woman for economic production, and, therefore, it is right that she should be supported by her husband.

The ground is taken that the human female is not economically independent, that she is fed by the male of her species. In denial of this, it is first alleged that she is economically independent,—that she does support herself by her own industry in the house. It being shown that there is no relation between the economic status of women and the labor she performs in the home, it is then alleged that not as house servant, but as mother, does woman earn her living. It being shown that the economic status of women bears no relation to her motherhood, either in quantity or quality, it is then alleged that motherhood renders a women unfit for economic production, and that, therefore, it is right that she be supported by her husband. Before going farther, let us seize upon this admission,—that she is supported by her husband.

Without going into either the ethics or the necessities of the case, we have reached so much common ground: the female of genus homo is supported by the male. Whereas, in other species of animals, male and female alike graze and browse, hunt and kill, climb, swim, dig, run, and fly for their livings, in our species the female does not seek her own living in the specific activities of our race, but is fed by the male.

Now as to the alleged necessity. Because of her maternal duties, the human female is said to be unable to get her own living. As the maternal duties of other females do not unfit them for getting their own living and also the livings of their young, it would seem that the human maternal duties require the segregation of the entire energies of the mother to the service of the child during her entire adult life, or so large a proportion of them that not enough remains to devote to the individual interests of the mother.

Such a condition, did it exist, would of course excuse and justify the pitiful dependence of the human female, and her support by the male. As the queen bee, modified entirely to maternity, is supported, not by the male, to be sure, but by her co-workers, the "old maids," the barren working bees, who labor so patiently and lovingly in their branch of the maternal duties of the hive, so would the human female, modified entirely to maternity, become unfit for any other exertion, and a helpless dependant.

Is this the condition of human motherhood? Does the human mother, by her motherhood, thereby lose control of brain and body, lose power and skill and desire for any other work? Do we see before us the human race, with all its females segregated entirely to the uses of motherhood, consecrated, set apart, specially developed, spending every power of their nature on the service of their children?

We do not. We see the human mother worked far harder than a mare, laboring her life long in the service, not of her children only, but of men; husbands, brothers, fathers, whatever male relatives she has; for mother and sister also; for the church a little, if she is allowed; for society, if she is able; for charity and education and reform,—working in many ways that are not the ways of motherhood.

It is not motherhood that keeps the housewife on her feet from dawn till dark; it is house service, not child service. Women work longer and harder than most men, and not solely in maternal duties. The savage mother carries the burdens, and does all menial service for the family tribe. The peasant mother toils in the fields, and the workingman's wife in the home. Many mothers, even now, are wage earners for the family, as well as bearers and rearers of it. And the women who are not so occupied, the women who belong to rich men,—here perhaps is the exhaustive devotion to maternity which is supposed to justify an admitted economic dependence. But we do not find it even among these. Women of ease and wealth provide for their children better care than the poor woman can; but they do not spend more time upon it themselves, nor more care and effort. They have other occupation.

In spite of her supposed segregation to maternal duties, the human female, the world over, works at extra-maternal duties for hours enough to provide her with an independent living, and then is denied independence on the ground that motherhood prevents her working! . . .

The working power of the mother has always been a prominent factor in human life. She is the worker *par excellence*, but her work is not such as to affect her economic status. Her living, all that she gets,—food, clothing, ornaments, amusements, luxuries,—these bear no relation to her power to produce wealth, to her services in the house, or to her motherhood. These things bear relation only to the man she marries, the man she depends on,—to how much he has and how much he is willing to give her. The women whose splendid extravagance dazzles the world, whose economic goods are the greatest, are often neither houseworkers nor mothers, but simply the women who hold most power over the men who have the most money. The female of genus homo is economically dependent on the male. He is her food supply. . . .

It should be clear to any one accustomed to the working of biological laws that all the tendencies of a living organism are progressive in their development, and are held in check by the interaction of their several forces. Each living form, with its dominant characteristics, represents a balance of power, a sort of compromise. The size of earth's primeval monsters was limited by the tensile strength of their material. Sea monsters can be bigger, because the medium in which they move offers more support. Birds must be smaller for the opposite reason. The cow requires many stomachs of a liberal size, because her food is of low nutritive value; and she must eat large quantities to keep her machine going. The size of arboreal animals, such as monkeys or squirrels, is limited by the nature of their habitat: creatures that live in trees cannot be so big as creatures that live on the ground. Every quality of every creature is relative to its condition, and tends to increase or decrease accordingly; and each quality tends to increase in proportion to its use, and to decrease in proportion to its disuse. Primitive man and his female were animals, like other animals. They were strong, fierce, lively beasts; and she was as nimble and ferocious as he, save for the added belligerence of the males in their sex-competition. In this competition, he, like the other male creatures, fought savagely with his hairy rivals; and she, like the other female creatures, complacently viewed their struggles, and mated with the victor. At other times she ran about in the forest, and helped herself to what there was to eat as freely as he did.

There seems to have come a time when it occurred to the dawning intelligence of this amiable savage that it was cheaper and easier to fight a little female, and have it done with, than to fight a big male every time. So he instituted the custom of enslaving the female; and she, losing freedom, could no longer get her own food nor that of her young. The mother ape, with her maternal function well fulfilled, flees leaping through the forest,—plucks her fruit and nuts, keeps up with the movement of the tribe, her young one on her back or held in one strong arm. But the mother woman, enslaved, could not do this. Then man, the father, found that slavery had its obligations: he must care for what he forbade to care for itself, else it died on his hands. So he slowly and reluctantly shouldered the duties of his new position. He began to feed her, and not only that, but to express in his own person the thwarted uses of maternity: he had to feed the children, too. It seems a simple arrangement. When we have thought of it at all, we have thought of it with admiration. The naturalist defends it on the ground of advantage to the species through the freeing of the mother from all other cares and confining her unreservedly to the duties of maternity. The poet and novelist, the painter and sculptor, the priest and teacher, have all extolled this lovely relation. It remains for the sociologist, from a biological point of view, to note its effects on the constitution of the human race, both in the individual and in society.

When man began to feed and defend women, she ceased proportionately to feed and defend herself. When he stood between her and her physical environment, she ceased proportionately to feel the influence of that environment and respond to it. When he became her immediate and all-important environment, she began proportionately to respond to this new influence, and to be modified accordingly. In a free state, speed was of as great advantage to the female as to the male, both in enabling her to catch prey and in preventing her from being caught by enemies; but, in her new condition, speed was a disadvantage. She was not allowed to do the catching, and it profited her to be caught by her new master. Free creatures, getting their own food and maintaining their own lives, develop an active capacity for attaining their ends. Parasitic creatures, whose living is obtained by the exertions of others, develop powers of absorption and of tenacity,—the powers by which they profit most. The human female was cut off from the direct action of natural selection, that mighty force which heretofore had acted on male and female alike with inexorable and beneficial effect, developing strength, developing skill, developing endurance, developing courage,—in a word, developing species. She now met the influence of natural selection acting indirectly through the male, and developing, of course, the faculties required to secure and obtain a hold on him. Needless to state that these faculties were those of sex-attraction, the one power that has made him cheerfully maintain, in what luxury he could, the being in whom he delighted. For many, many centuries she had no other hold, no other assurance of being fed. The young girl had a prospective value, and was maintained for what should follow; but the old woman, in more primitive times, had but a poor hold on life. She who could best please her lord was the favorite slave or favorite wife, and she obtained the best economic conditions.

With the growth of civilization, we have gradually crystallized into law the visible necessity for feeding the helpless female; and even old women are maintained by their male relatives with a comfortable assurance. But to this day—save, indeed, for the increasing army of women wage-earners, who are changing the face of the world by their steady advance toward economic independence—the personal profit of women bears but too close a relation to their power to win and hold the other sex. From the odalisque with the most bracelets to the débutante with the most bouquets, the relation still holds good,—woman's economic profit comes through the power of sex-attraction.

When we confront this fact boldly and plainly in the open market of vice, we are sick with horror. When we see the same economic relation made permanent, established by law, sanctioned and sanctified by religion, covered with flowers and incense and all accumulated sentiment, we think it innocent, lovely, and right. The transient trade we think evil. The bargain for life we think good. But the biological effect remains the same. In both cases the female gets her food from the male by virtue of her sex-relationship to him. In both cases, perhaps even more in marriage because of its perfect acceptance of the situation, the female of genus homo, still living under natural law, is inexorably modified to sex in an increasing degree.

Followed in specific detail, the action of the changed environment upon women has been in given instances as follows: In the matter of mere passive surroundings she has been immediately restricted in her range. This one factor has an immense effect on man and animal alike. An absolutely uniform environment, one shape, one size, one color, one sound, would render life, if any life could be, one helpless, changeless thing. As the environment increases and varies, the development of the creature must increase and vary with it; for he acquires knowledge and power, as the material for knowledge and the need for power appear. In migratory species the female is free to acquire the same knowledge as the male by the same means, the same development by the same experiences. The human female has been restricted in range from the earliest beginning. Even among savages, she has a much more restricted knowledge of the land she lives in. She moves with the camp, of course, and follows her primitive industries in its vicinity; but the war-path and the hunt are the man's. He has a far larger habitat. The life of the female savage is freedom itself, however, compared with the increasing constriction of custom closing in upon the woman, as civilization advanced, like the iron torture chamber of romance. Its culmination is expressed in the proverb: "A woman should leave her home but three times,—when she is christened, when she is married, and when she is buried." Or this: "The woman, the cat, and the chimney should never leave the house." The absolutely stationary female and the wide-ranging male are distinctly human institutions, after we leave behind us such low forms of life as the gypsy moth, whose female seldom moves more than a few feet from the pupa moth. She has aborted wings, and cannot fly. She waits humbly for the winged male, lays her myriad eggs, and dies,— a fine instance of modification to sex.

To reduce so largely the mere area of environment is a great check to race-development; but it is not to be compared in its effects with the reduction in voluntary activity to which the human female has been subjected. Her restricted impression, her confinement to the four walls of the home, have done great execution, of course, in limiting her ideas, her information, her thought-processes, and power of judgment; and in giving a disproportionate prominence and intensity to the few things she knows about; but this is innocent in action compared with her restricted expression, the denial of freedom to act. A living organism is modified far less through the action of external circumstances upon it and its reaction thereto, than through the effect of its own exertions. Skin may be thickened gradually by exposure to the weather; but it is thickened far more quickly by being rubbed against something, as the handle of an oar or of a broom. To be surrounded by beautiful things has much influence upon the human creature: to make beautiful things has more. To live among beautiful surroundings and make ugly things is more directly lowering than to live among ugly surroundings and make beautiful things. What we do modifies us more than what is done to us. The freedom of expression has been more restricted in women than the freedom of impression, if that be possible. Something of the world she lived in she has seen from her barred windows. Some air has come through the purdah's folds, some knowledge has filtered to her eager ears from the talk of men. Desdemona learned somewhat of Othello. Had she known more, she might have lived longer. But in the ever-growing human impulse to create, the power and will to make, to do, to express one's new spirit in new forms,—here she has been utterly debarred. She might work as she had worked from the beginning,—at the primitive labors of the household; but in the inevitable expansion of even those industries to professional levels we have striven to hold her back. To work with her own hands, for nothing, in direct body-service to her own family,—this has been permitted,—yes, compelled. But to be and do anything further from this she has been forbidden. Her labor has not only been limited in kinds, but in degree. Whatever she has been allowed to do must be done in private and alone, the first-hand industries of savage times.

Our growth in industry has been not only in kind, but in class. The baker is not in the same industrial grade with the house-cook, though both make bread. To specialize any form of labor is a step up: to organize it is another step. Specialization and organization are the basis of human progress, the organic methods of social life. They have been forbidden to women almost absolutely. The greatest and most beneficent change of this century is the progress of women in these two lines of advance. The effect of this check in industrial development, accompanied as it was by the

constant inheritance of increased racial power, has been to intensify the sensations and emotions of women, and to develop great activity in the lines allowed. The nervous energy that up to present memory has impelled women to labor incessantly at something, be it the veriest folly of fancy work, is one mark of this effect.

In religious development the same dead-line has held back the growth of women through all the races and ages. In dim early times she was sharer in the mysteries and rites; but, as religion developed, her place receded, until Paul commanded her to be silent in the churches. And she has been silent until today. Even now, with all the ground gained, we have but the beginnings—the slowly forced and disapproved beginnings—of religious equality for the sexes. In some nations, religion is held to be a masculine attribute exclusively, it being even questioned whether women have souls. An early Christian council settled that important question by vote, fortunately deciding that they had. In a church whose main strength has always been derived from the adherence of women, it would have been an uncomfortable reflection not to have allowed them souls. Ancient family worship ran in the male line. It was the son who kept the sacred grandfathers in due respect, and poured libations to their shades. When the woman married, she changed her ancestors, and had to worship her husband's progenitors instead of her own. This is why the Hindu and the Chinaman and many others of like stamp must have a son to keep them in countenance,—a deep-seated sex-prejudice, coming to slow extinction as women rise in economic importance.

It is painfully interesting to trace the gradual cumulative effect of these conditions upon women: first, the action of large natural laws, acting on her as they would act on any other animal; then the evolution of social customs and laws (with her position as the active cause), following the direction of mere physical forces, and adding heavily to them; then, with increasing civilization, the unbroken accumulation of precedent, burnt into each generation by the growing force of education, made lovely by art, holy by religion, desirable by habit; and, steadily acting from beneath, the unswerving pressure of economic necessity upon which the whole structure rested. These are strong modifying conditions, indeed.

The process would have been even more effective and far less painful but for one important circumstance. Heredity has no Salic law. Each girl-child inherits from her father a certain increasing percentage of human development, human power, human tendency; and each boy as well inherits from his mother the increasing percentage of sex-development, sex-power, sex-tendency. The action of heredity has been to equalize what every tendency of environment and education made to differ. This has saved us from such a female as the gypsy moth. It has held up the woman, and held down the man. It has set iron bounds to our absurd effort to make a race with one sex a million years behind the other. But it has added terribly to the pain and difficulty of human life,—a difficulty and a pain that should have taught us long since that we were living on wrong lines. Each woman born, re-humanized by the current of race activity carried on by her father and re-womanized by her traditional position, has had to live over again in her own person the same process of restriction, repression, denial; the smothering "no" which crushed down all her human desires to create, to discover, to learn, to express, to advance. Each woman has had, on the other hand, the same single avenue of expression and attainment; the same one way in which alone she might do what she could, get what she might. All other doors were shut, and this one always open; and the whole pressure of advancing humanity was upon her. No wonder that young Daniel in the apocryphal tale proclaimed: "The king is strong! Wine is strong! But women are stronger!"

To the young man confronting life the world lies wide. Such powers as he has he may use, must use. If he chooses wrong at first, he may choose again, and yet again. Not effective or successful in one channel, he may do better in another. The growing, varied needs of all mankind call on him for the varied service in which he finds his growth. What he wants to be, he may strive to be. What he wants to get, he may strive to get. Wealth, power, social distinction, fame,—what he wants he can try for.

To the young woman confronting life there is the same world beyond, there are the same human energies and human desires and ambition within. But all that she may wish to have, all that she may wish to do, must come through a single channel and a single choice. Wealth, power, social distinction, fame,—not only these, but home and happiness, reputation, ease and pleasure, her bread and butter,—all must come to her through a small gold ring. This is a heavy pressure. It has accumulated behind her through heredity, and continued about her through environment. It has been subtly trained into her through education, till she herself has come to think it a right condition, and pours its influence upon her daughter with increasing impetus. Is it any wonder that women are oversexed? But for the constant inheritance from the more human male, we should have been queen bees, indeed, long before this. But the daughter of the soldier and the sailor, of the artist, the inventor, the great merchant, has inherited in body and brain her share of his development in each generation, and so stayed somewhat human for all her femininity.

All morbid conditions tend to extinction. One check has always existed to our inordinate sex-development,—nature's ready relief, death. Carried to its furthest excess, the individual has died, the family has become extinct, the nation itself has perished, like Sodom and Gomorrah. Where one function is carried to unnatural excess, others are weakened, and the organism perishes. We are familiar with this in individual cases,—at least, the physician is. We can see it somewhat in the history of nations. From younger races, nearer savagery, nearer the healthful equality of pre-human creatures, has come each new start in history. Persia was older than Greece, and its highly differentiated sexuality had produced the inevitable result of enfeebling the racial qualities. The Greek commander stripped the rich robes and jewels from his Persian captives, and showed their unmanly feebleness to his men. "You have such bodies as these to fight for such plunder as this," he said. In the country, among peasant classes, there is much less sex-distinction than in cities, where wealth enables the women to live in absolute idleness; and even the men manifest the same characteristics. It is from the country and the lower classes that the fresh blood pours into the cities, to be weakened in its turn by the influence of this unnatural distinction until there is none left to replenish the nation.

The inevitable trend of human life is toward higher civilization; but, while that civilization is confined to one sex, it inevitably exaggerates sex-distinction, until the increasing evil of this condition is stronger than all the good of the civilization attained, and the nation falls. Civilization, be it understood, does not consist in the acquisition of luxuries. Social development is an organic development. A civilized State is one in which the citizens live in organic industrial relation. The more full, free, subtle, and easy that relation; the more perfect the differentiation of labor and exchange of product, with their correlative institutions,—the more perfect is that civilization. To eat, drink, sleep, and keep warm,—these are common to all animals, whether the animal couches in a bed of leaves or one of eiderdown, sleeps in the sun to avoid the wind or builds a furnace-heated house, lies in wait for game or orders a dinner at a hotel. These are but individual animal processes. Whether one lays an egg or a million eggs, whether one bears a cub, a kitten, or a baby, whether one broods its chickens, guards its litter, or tends a nursery full of children, these are but individual animal processes. But to serve each other more and more widely; to live only by such service; to develop special functions, so that we depend for our living on society's return for services that can be of no direct use to ourselves,—this is civilization, our human glory and race-distinction.

All this human progress has been accomplished by men. Women have been left behind, outside, below, having no social relation whatever, merely the sex-relation, whereby they lived. Let us bear in mind that all the tender ties of family are ties of blood, of sex-relationship. A friend, a comrade, a partner,—this is a human relative. Father, mother, son, daughter, sister, brother, husband, wife,—these are sex-relatives. Blood is thicker than water, we say. True. But ties of blood are not those that ring the world with the succeeding waves of progressive religion, art, science, commerce, education, all that makes us human. Man is the human creature. Woman has been checked, starved, aborted in human growth; and the swelling forces of race-development have been driven back in each generation to work in her through sex-functions alone....

★★★

The facts stated in the foregoing chapters are familiar and undeniable, the argument seems clear; yet the mind reacts violently from the conclusions it is forced to admit, and tries to find relief in the commonplace conditions of every-day life....

It is equally possible for a race, a nation, a class, to become accustomed to most disadvantageous conditions, and fail to notice them. Take, as an individual instance, the wearing of corsets by women. Put a corset, even a loose one, on a vigorous man or woman who never wore one, and there is intense discomfort, and a vivid consciousness thereof. The healthy muscles of the trunk resent the pressure, the action of the whole body is checked in the middle, the stomach is choked, the process of digestion interfered with; and the victim says, "How can you bear such a thing?"

But the person habitually wearing a corset does not feel these evils. They exist, assuredly, the facts are there, the body is not deceived; but the nerves have become accustomed to these disagreeable sensations, and no longer respond to them. The person "does not feel it." In fact, the wearer becomes so used to the sensations that, when they are removed,—with the corset,—there is a distinct sense of loss and discomfort. The heavy folds of the cravat, stock, and neckcloth of earlier men's fashions, the heavy horsehair peruke, the stiff high collar of today, the kind of shoes we wear,—these are perfectly familiar instances of the force of habit in the individual.

This is equally true of racial habits. That a king should rule because he was born, passed unquestioned for thousands of years. That the eldest son should inherit the titles and estates was a similar phenomenon as little questioned. That a debtor should be imprisoned, and so entirely prevented from paying his debts, was common law. So glaring an evil as chattel slavery was an unchallenged social institution from earliest history to our own day among the most civilized nations of the earth. Christ himself let it pass unnoticed. The hideous injustice of Christianity to the Jew attracted no attention through many centuries. That the serf went with the soil, and was owned by the lord thereof, was one of the foundations of society in the Middle Ages.... When a condition exists among us which began in those unrecorded ages back of tradition even, which obtains in varying degree among every people on earth, and which begins to act upon the individual at birth, it would be a miracle past all belief if people should notice it. The sexuo-economic relation is such a condition. It began in primeval savagery. It exists in all nations. Each boy and girl is born into it, trained into it, and has to live in it. The world's progress in matters like these is attained by a slow and painful process, but one which works to good ends....

DISCUSSION QUESTIONS

1. How is the economic status of human beings correlated with the sex-relation?

2. At one point in her essay, Charlotte Perkins Gilman suggests that women are seen as economically dependent both individually and collectively. Explain what she means by this.

3. Explain Gilman's statement that there is no relationship between women's status and their domestic labor, "unless an inverse one." What do you think she means by this?

4. In today's society, are there any "disadvantageous conditions" for women or men?

5. Place this piece in its historical context. What was happening when Gilman wrote this piece (1898)? What would she write now?

6. Review the glossary definitions for sex and gender in the back of this volume. Now think about how Gilman used the terms in her writing. What does this tell you about the thinking about gender and sexuality during the time the article was written?

INTERNET RESOURCES

Suggested Web URLs for Further Study

http://womenwriters.net/domesticgoddess/gilman2.html
This is a site about Gilman herself, with other works and many other links to women writers.

http://web.cortland.edu/gilman/
This site is a society for admirers of Gilman and other female writers that contains other links, information about conferences, and so on.

Virtual Society

For further information on this subject, including links to relevant websites, go to the Wadsworth Sociology homepage at
http://www.cengage.com/sociology.

13

The Power Elite

C. WRIGHT MILLS

The powers of ordinary men are circumscribed by the everyday worlds in which they live, yet even in these rounds of job, family, and neighborhood they often seem driven by forces they can neither understand nor govern. "Great changes" are beyond their control, but affect their conduct and outlook nonetheless. The very framework of modern society confines them to projects not their own, but from every side, such changes now press upon the men and women of the mass society, who accordingly feel that they are without purpose in an epoch in which they are without power.

But not all men are in this sense ordinary. As the means of information and of power are centralized, some men come to occupy positions in American society from which they can look down upon, so to speak, and by their decisions mightily affect, the everyday worlds of ordinary men and women. They are not made by their jobs; they set up and break down jobs for thousands of others; they are not confined by simple family responsibilities; they can escape. They may live in many hotels and houses, but they are bound by no one community. They need not merely "meet the demands of the day and hour"; in some part, they create these demands, and cause others to meet them. Whether or not they profess their power, their technical and political experience of it far transcends that of the underlying population. What Jacob Burckhardt said of "great men," most Americans might well say of their elite: "They are all that we are not."

The power elite is composed of men whose positions enable them to transcend the ordinary environments of ordinary men and women; they are in positions to make decisions having major consequences. Whether they do or do not make such decisions is less important than the fact that they do occupy such pivotal positions: Their failure to act, their failure to make decisions, is itself an act that is often of greater consequence than the decisions they do make. For they are in command

SOURCE: From *The Power Elite, New Edition* by C. Wright Mills, © 1956, 2000 by Oxford University Press, Inc. Used by permission of Oxford University Press, Inc.

of the major hierarchies and organizations of modern society. They rule the big corporations. They run the machinery of the state and claim its prerogatives. They direct the military establishment. They occupy the strategic command posts of the social structure, in which are now centered the effective means of the power and the wealth and the celebrity which they enjoy. The power elite are not solitary rulers. Advisers and consultants, spokesmen and opinion-makers are often the captains of their higher thought and decision. Immediately below the elite are the professional politicians of the middle levels of power, in the Congress and in the pressure groups, as well as among the new and old upper classes of town and city and region. Mingling with them, in curious ways which we shall explore, are those professional celebrities who live by being continually displayed but are never, so long as they remain celebrities, displayed enough. If such celebrities are not at the head of any dominating hierarchy, they do often have the power to distract the attention of the public or afford sensations to the masses, or, more directly, to gain the ear of those who do occupy positions of direct power. More or less unattached, as critics of morality and technicians of power, as spokesmen of God and creators of mass sensibility, such celebrities and consultants are part of the immediate scene in which the drama of the elite is enacted. But that drama itself is centered in the command posts of the major institutional hierarchies.

The truth about the nature and the power of the elite is not some secret which men of affairs know but will not tell. Such men hold quite various theories about their own roles in the sequence of event and decision. Often they are uncertain about their roles, and even more often they allow their fears and their hopes to affect their assessment of their own power. No matter how great their actual power, they tend to be less acutely aware of it than of the resistances of others to its use. Moreover, most American men of affairs have learned well the rhetoric of public relations, in some cases even to the point of using it when they are alone, and thus coming to believe it. The personal awareness of the actors is only one of the several sources one must examine in order to understand the higher circles. Yet many who believe that there is no elite, or at any rate none of any consequence, rest their argument upon what men of affairs believe about themselves, or at least assert in public.

There is, however, another view: Those who feel, even if vaguely, that a compact and powerful elite of great importance does now prevail in America often base that feeling upon the historical trend of our time. They have felt, for example, the domination of the military event, and from this they infer that generals and admirals, as well as other men of decision influenced by them, must be enormously powerful. They hear that the Congress has again abdicated to a handful of men decisions clearly related to the issue of war or peace. They know that the bomb was dropped over Japan in the name of the United States of America, although they were at no time consulted about the matter. They feel that they live in a time of big decisions; they know that they are not making any. Accordingly, as they consider the present as history, they infer that at its center, making decisions or failing to make them, there must be an elite of power.

On the one hand, those who share this feeling about big historical events assume that there is an elite and that its power is great. On the other hand, those who listen carefully to the reports of men apparently involved in the great decisions often do not believe that there is an elite whose powers are of decisive consequence.

Both views must be taken into account, but neither is adequate. The way to understand the power of the American elite lies neither solely in recognizing the historic scale of events nor in accepting the personal awareness reported by men of apparent decision. Behind such men and behind the events of history, linking the two, are the major institutions of modern society. These hierarchies of state and corporation and army constitute the means of power; as such they are now of a consequence not before equaled in human history—and at their summits, there are now those command posts of modern society which offer us the sociological key to an understanding of the role of the higher circles in America.

Within American society, major national power now resides in the economic, the political, and the military domains. Other institutions seem off to the side of modern history, and, on occasion, duly subordinated to these. No family is as directly powerful in national affairs as any major corporation; no church is as directly powerful in the external biographies of young men in America today as the military establishment; no college is as powerful in the shaping of momentous events as the National Security Council. Religious, educational, and family institutions are not autonomous centers of national power; on the contrary, these decentralized areas are increasingly shaped by the big three, in which developments of decisive and immediate consequence now occur.

Families and churches and schools adapt to modern life; governments and armies and corporations shape it; and, as they do so, they turn these lesser institutions into means for their ends. Religious institutions provide chaplains to the armed forces where they are used as a means of increasing the effectiveness of its morale to kill. Schools select and train men for their jobs in corporations and their specialized tasks in the armed forces. The extended family has, of course, long been broken up by the industrial revolution, and now the son and the father are removed from the family, by compulsion if need be, whenever the army of the state sends out the call. And the symbols of all these lesser institutions are used to legitimate the power and the decisions of the big three.

The life-fate of the modern individual depends not only upon the family into which he was born or which he enters by marriage, but increasingly upon the corporation in which he spends the most alert hours of his best years; not only upon the school where he is educated as a child and adolescent, but also upon the state which touches him throughout his life; not only upon the church in which on occasion he hears the word of God, but also upon the army in which he is disciplined.

If the centralized state could not rely upon the inculcation of nationalist loyalties in public and private schools, its leaders would promptly seek to modify the decentralized educational system. If the bankruptcy rate among the top 500 corporations were as high as the general divorce rate among the 37 million married couples, there would be economic catastrophe on an international scale. If members of armies gave to them no more of their lives than do

believers to the churches to which they belong, there would be a military crisis.

Within each of the big three, the typical institutional unit has become enlarged, has become administrative, and, in the power of its decisions, has become centralized. Behind these developments there is a fabulous technology, for as institutions, they have incorporated this technology and guide it, even as it shapes and paces their developments.

The economy—once a great scatter of small productive units in autonomous balance—has become dominated by two or three hundred giant corporations, administratively and politically interrelated, which together hold the keys to economic decisions.

The political order, once a decentralized set of several dozen states with a weak spinal cord, has become a centralized, executive establishment which has taken up into itself many powers previously scattered, and now enters into each and every cranny of the social structure.

The military order, once a slim establishment in a context of distrust fed by state militia, has become the largest and most expensive feature of government, and, although well-versed in smiling public relations, now has all the grim and clumsy efficiency of a sprawling bureaucratic domain. In each of these institutional areas, the means of power at the disposal of decision makers have increased enormously; their central executive powers have been enhanced; within each of them modern administrative routines have been elaborated and tightened up.

As each of these domains becomes enlarged and centralized, the consequences of its activities become greater, and its traffic with the others increases. The decisions of a handful of corporations bear upon military and political as well as upon economic developments around the world. The decisions of the military establishment rest upon and grievously affect political life as well as the very level of economic activity. The decisions made within the political domain determine economic activities and military programs. There is no longer, on the one hand, an economy, and, on the other hand, a political order containing a military establishment unimportant to politics and to money-making. There is a political economy linked, in a thousand ways, with military institutions and decisions. On each side of the world-split running through central Europe and around the Asiatic rimlands, there is an ever-increasing interlocking of economic, military, and political structures. If there is government intervention in the corporate economy, so is there corporate intervention in the governmental process. In the structural sense, this triangle of power is the source of the interlocking directorate that is most important for the historical structure of the present.

The fact of the interlocking is clearly revealed at each of the points of crisis of modern capitalist society—slump, war, and boom. In each, men of decision are led to an awareness of the interdependence of the major institutional orders. In the nineteenth century, when the scale of all institutions was smaller, their liberal integration was achieved in the automatic economy, by an autonomous play of market forces, and in the automatic political domain, by the bargain and the vote. It was then assumed that out of the imbalance and friction that followed the limited decisions then possible a new equilibrium would in due course emerge. That can no longer be assumed, and it is not assumed by the men at the top of each of the three dominant hierarchies.

For given the scope of their consequences, decisions—and indecisions—any one of these ramify into the others, and hence top decisions tend either to become coordinated or to lead to a commanding indecision. It has not always been like this. When numerous small entrepreneurs made up the economy, for example, many of them could fail and the consequences still remain local; political and military authorities did not intervene. But now, given political expectations and military commitments, can they afford to allow key units of the private corporate economy break down in [a] slump? Increasingly, they do intervene in economic affairs, and as they do so, the controlling decisions in each order are inspected by agents of the other two, and economic, military, and political structures are interlocked.

At the pinnacle of each of the three enlarged and centralized domains, there have arisen those higher circles which make up the economic, the political, and the military elites. At the top of the economy, among the corporate rich, there are the chief executives; at the top of the political order, the members of the political directorate; at the top of the military establishment, the elite of soldier-statesmen clustered in and around the Joint Chiefs of Staff and the upper echelon. As each of these domains has coincided with the others, as decisions tend to become total in their consequence, the leading men in each of the three domains of power—the warlords, the corporation chieftains, the political directorate—tend to come together, to form the power elite of America.

The higher circles in and around these command posts are often thought of in terms of what their members possess: They have a greater share than other people of the things and experiences that are most highly valued. From this point of view, the elite are simply those who have the most of what there is to have, which is generally held to include money, power, and prestige—as well as all the ways of life to which these lead. But the elite are not simply those who have the most, for they could not "have the most" were it not for their positions in the great institutions. For such institutions are the necessary bases of power, of wealth, and of prestige, and at the same time, the chief means of exercising power, of acquiring and retaining wealth, and of cashing in the higher claims for prestige.

By the powerful we mean, of course, those who are able to realize their will, even if others resist it. No one, accordingly, can be truly powerful unless he has access to the command of major institutions, for it is over these institutional means of power that the truly powerful are, in the first instance, powerful. Higher politicians and key officials of government command such institutional power; so do admirals and generals, and so do the major owners and executives of the larger corporations. Not all power, it is true, is anchored in and exercised by means of such institutions, but only within and through them can power be more or less continuous and important.

Wealth also is acquired and held in and through institutions. The pyramid of wealth cannot be understood merely in terms of the very rich; for the great inheriting families, as we shall see, are now supplemented by the coorporate institutions of modern society: Every one of the very rich families has been and is closely connected—always legally and frequently managerially as well—with one of the multimillion-dollar corporations. The modern corporation is the prime source of wealth, but, in latter-day capitalism, the political apparatus also opens and closes many avenues to wealth. The amount as well as the source of income, the power over consumer's goods as well as over productive capital, are determined by position within the political economy. If our interest in the very rich goes beyond their lavish or their miserly consumption, we must examine their relations to modern forms of corporate property as well as to the state; for such relations now determine the chances of men to secure big property and to receive high income.

Great prestige increasingly follows the major institutional units of the social structure. It is obvious that prestige depends, often quite decisively, upon access to the publicity machines that are now a central and normal feature of all the big institutions of modern America. Moreover, one feature of these hierarchies of corporation, state, and military establishment is that their top positions are increasingly interchangeable. One result of this is the accumulative nature of prestige. Claims for prestige, for example, may be initially based on military roles, then expressed in and augmented by an educational institution run by corporate executives, and cashed in, finally, in the political order, where, for General Eisenhower and those he represents, power and prestige finally meet at the very peak. Like wealth and power, prestige tends to be cumulative: The more of it you have, the more you can get. These values also tend to be translatable into one another: The wealthy find it easier than the poor to gain power; those with status find it easier than those without it to control opportunities for wealth.

If we took the one-hundred most powerful men in America, the one-hundred wealthiest, and the one-hundred most celebrated away from the institutional positions they now occupy, away from their resources of men and women and money, away from the media of mass communication that are now focused upon them—then they would be powerless and poor and uncelebrated. For power is not of a man. Wealth does not center in the person of the wealthy. Celebrity is not inherent in any personality. To be celebrated, to be wealthy, to have

power requires access to major institutions, for the institutional positions men occupy determine in large part their chances to have and to hold these valued experiences.

The people of the higher circles may also be conceived as members of a top social stratum, as a set of groups whose members know one another, see one another socially and at business, and so, in making decisions, take one another into account. The elite, according to this conception, feel themselves to be, and are felt by others to be, the inner circle of "the upper social classes." They form a more or less compact social and psychological entity; they have become self-conscious members of a social class. People are either accepted into this class or they are not, and there is a qualitative split, rather than merely a numerical scale, separating them from those who are not elite. They are more or less aware of themselves as a social class and they behave toward one another differently from the way they do toward members of other classes. They accept one another, understand one another, marry one another, tend to work and to think if not together at least alike.

Now, we do not want by our definition to prejudge whether the elite of the command posts are conscious members of such a socially recognized class, or whether considerable proportions of the elite derive from such a clear and distinct class. These are matters to be investigated. Yet in order to be able to recognize what we intend to investigate, we must note something that all biographies and memoirs of the wealthy and the powerful and the eminent make clear: No matter what else they may be, the people of these higher circles are involved in a set of overlapping "crowds" and intricately connected "cliques." There is a kind of mutual attraction among those who "sit on the same terrace"—although this often becomes clear to them, as well as to others, only at the point at which they feel the need to draw the line; only when, in their common defense, they come to understand what they have in common, and so close their ranks against outsiders.

The idea of such ruling stratum implies that most of its members have similar social origins, that throughout their lives they maintain a network of internal connections, and that to some degree there is an interchangeability of position between the various hierarchies of money and power and celebrity. We must, of course, note at once that if such an elite stratum does exist, its social visibility and its form, for very solid historical reasons, are quite different from those of the noble cousin-hoods that once ruled various European nations.

That American society has never passed through a feudal epoch is of decisive importance to the nature of the American elite, as well as to American society as a historic whole. For it means that no nobility or aristocracy, established before the capitalist era, has stood in tense opposition to the higher bourgeoisie. It means that this bourgeoisie has monopolized not only wealth but prestige and power as well. It means that no set of noble families has commanded the top positions and monopolized the values that are generally held in high esteem; and certainly that no set has done so explicitly by inherited right. It means that no high church dignitaries or court nobilities, no

entrenched landlords with honorific accouterments, no monopolists of high army posts have opposed the enriched bourgeoisie and in the name of birth and prerogative successfully resisted its self-making. But this does *not* mean that there are no upper strata in the United States. That they emerged from a "middle class" that had no recognized aristocratic superiors does not mean they remained middle class when enormous increases in wealth made their own superiority possible. Their origins and their newness may have made the upper strata less visible in America than elsewhere. But in America today there are in fact tiers and ranges of wealth and power of which people in the middle and lower ranks know very little and may not even dream. There are families who, in their well-being, are quite insulated from the economic jolts and lurches felt by the merely prosperous and those farther down the scale. There are also men of power who in quite small groups make decisions of enormous consequence for the underlying population....

DISCUSSION QUESTIONS

1. C. Wright Mills sees the power of certain institutions as having increased over time. What evidence do you see of the power of these institutions in your life?

2. The power elite, although composed of individuals, wields its influence because of its institutional control. How do these institutions

interlock in the current period, and what consequences does this have for ordinary men and women?

3. What role does Mills see celebrities having with regard to the power elite? What evidence do you see of this in contemporary times?

INTERNET RESOURCES

Suggested Web URLs for Further Study

http://www.socialstudieshelp.com/APGOV_Power_Elite.htm
This website features an essay on *The Power Elite* and provides specific examples that relate to Mills's theories.

http://www.angelfire.com/or/sociologyshop/CWM.html

This web page has links to other articles by Mills and other essays commenting on Mills and his works.

Virtual Society

For further information on this subject, including links to relevant websites, go to the Wadsworth Sociology homepage at *http://www.cengage.com/sociology*.

14

Savage Inequalities

JONATHAN KOZOL

"East of anywhere," writes a reporter for the *St. Louis Post-Dispatch*, "often evokes the other side of the tracks. But, for a first-time visitor suddenly deposited on its eerily empty streets, East St. Louis might suggest another world." The city, which is 98 percent black, has no obstetric services, no regular trash collection, and few jobs. Nearly a third of its families live on less than $7,500 a year: 75 percent of its population lives on welfare of some form. The U.S. Department of Housing and Urban Development describes it as "the most distressed small city in America."

Only three of the 13 buildings on Missouri Avenue, one of the city's major thoroughfares, are occupied. A 13-story office building, tallest in the city, has been boarded up. Outside, on the sidewalk, a pile of garbage fills a ten-foot crater.

The city, which by night and day is clouded by the fumes that pour from vents and smokestacks at the Pfizer and Monsanto chemical plants, has one of the highest rates of child asthma in America.

It is, according to a teacher at Southern Illinois University, "a repository for a nonwhite population that is now regarded as expendable." The *Post-Dispatch* describes it as "America's Soweto."

Fiscal shortages have forced the layoff of 1,170 of the city's 1,400 employees in the past 12 years. The city, which is often unable to buy heating fuel or toilet paper for the city hall, recently announced that it might have to cashier all but 10 percent of the remaining work force of 230. In 1989 the mayor announced that he might need to sell the city hall and all six fire stations to raise needed cash. Last year the plan had to be scrapped after the city lost its city hall in a court judgment to a creditor. East St. Louis is mortgaged into the next century but has the highest property-tax rate in the state....

The dangers of exposure to raw sewage, which backs up repeatedly into the homes of residents in East St. Louis, were first noticed, in the spring of 1989, at a

SOURCE: From *Savage Inequalities* by Jonathan Kozol, © 1991 by Jonathan Kozol. Used by permission of Crown Publishers, a division of Random House, Inc.

public housing project, Villa Griffin. Raw sewage, says the *Post-Dispatch*, over-flowed into a playground just behind the housing project, which is home to 187 children, "forming an oozing lake of … tainted water." … A St. Louis health official voices her dismay that children live with waste in their backyards. "The development of working sewage systems made cities livable a hundred years ago," she notes. "Sewage systems separate us from the Third World." …

The sewage, which is flowing from collapsed pipes and dysfunctional pumping stations, has also flooded basements all over the city. The city's vacuum truck, which uses water and suction to unclog the city's sewers, cannot be used because it needs $5,000 in repairs. Even when it works, it sometimes can't be used because there isn't money to hire drivers. A single engineer now does the work that 14 others did before they were laid off. By April the pool of overflow behind the Villa Griffin project has expanded into a lagoon of sewage. Two million gallons of raw sewage lie outside the children's homes.…

Sister Julia Huiskamp meets me on King Boulevard and drives me to the Griffin homes.

As we ride past blocks and blocks of skeletal structures, some of which are still inhabited, she slows the car repeatedly at railroad crossings. A seemingly endless railroad train rolls past us to the right. On the left: a blackened lot where garbage has been burning. Next to the burning garbage is a row of 12 white cabins, charred by fire. Next: a lot that holds a heap of auto tires and a mountain of tin cans. More burnt houses. More trash fires. The train moves almost imperceptibly across the flatness of the land.

Fifty years old, and wearing a blue suit, white blouse, and blue headcover, Sister Julia points to the nicest house in sight. The sign on the front reads MOTEL. "It's a whorehouse," Sister Julia says.

When she slows the car beside a group of teenage boys, one of them steps out toward the car, then backs away as she is recognized.

The 99 units of the Villa Griffin homes—two-story structures, brick on the first floor, yellow wood above—form one border of a recessed park and playground that were filled with fecal matter last year when the sewage mains exploded. The sewage is gone now and the grass is very green and looks inviting. When nine-year-old Serena and her seven-year-old brother take me for a walk, however, I discover that our shoes sink into what is still a sewage marsh. An inch-deep residue of fouled water still remains.

Serena's brother is a handsome, joyous little boy, but troublingly thin. Three other children join us as we walk along the marsh: Smokey, who is nine years old but cannot yet tell time; Mickey, who is seven; and a tiny child with a ponytail and big brown eyes who talks a constant stream of words that I can't always understand.

"Hush, Little Sister," says Serena. I ask for her name, but "Little Sister" is the only name the children seem to know.

"There go my cousins," Smokey says, pointing to two teen-age girls above us on the hill.

The day is warm, although we're only in the second week of March; several dogs and cats are playing by the edges of the marsh. "It's a lot of squirrels here," says Smokey. "There go one!" "This here squirrel is a friend of mine," says Little Sister.

None of the children can tell me the approximate time that school begins. One says five o'clock. One says six. Another says that school begins at noon.

When I ask what song they sing after the flag pledge, one says, "Jingle Bells."

Smokey cannot decide if he is in the second or third grade.

Seven-year-old Mickey sucks his thumb during the walk.

The children regale me with a chilling story as we stand beside the marsh. Smokey says his sister was raped and murdered and then dumped behind his school. Other children add more details: Smokey's sister was 11 years old. She was beaten with a brick until she died. The murder was committed by a man who knew her mother.

The narrative begins when, without warning, Smokey says, "My sister has got killed."

"She was my best friend," Serena says.

"They had beat her in the head and raped her," Smokey says.

"She was hollering out loud," says Little Sister.

I ask them when it happened. Smokey says, "Last year." Serena then corrects him and she says, "Last week."

"It scared me because I had to cry," says Little Sister.

"The police arrested one man but they didn't catch the other," Smokey says.

Serena says, "He was some kin to her."

But Smokey objects, "He weren't no kin to me. He was my momma's friend." "Her face was busted," Little Sister says.

Serena describes this sequence of events: "They told her go behind the school. They'll give her a quarter if she do. Then they knock her down and told her not to tell what they had did."

I ask, "Why did they kill her?"

"They was scared that she would tell," Serena says.

"One is in jail," says Smokey. "They cain't find the other."

"Instead of raping little bitty children, they should find themselves a wife," says Little Sister.

"I hope," Serena says, "her spirit will come back and get that man."

"And *kill* that man," says Little Sister.

"Give her another chance to live," Serena says. "My teacher came to the funeral," says Smokey.

"When a little child dies, my momma say a star go straight to Heaven," says Serena.

"My grandma was murdered," Mickey says out of the blue. "Somebody shot two bullets in her head."

I ask him, "Is she really dead?"

"She dead all right," says Mickey. "She was layin' there, just dead."

"I love my friends," Serena says. "I don't care if they no kin to me. I care for them. I hope his mother have another baby. Name her for my friend that's dead."

"I have a cat with three legs," Smokey says. "Snakes hate rabbits," Mickey says, again for no apparent reason.

"Cats hate fishes," Little Sister says. "It's a lot of hate," says Smokey.

Later, at the mission, Sister Julia tells me this: "The Jefferson School, which they attend, is a decrepit hulk. Next to it is a modern school, erected two years ago, which was to have replaced the one that they attend. But the construction was not done correctly. The roof is too heavy for the walls, and the entire structure has begun to sink. It can't be occupied. Smokey's sister was raped and murdered and dumped between the old school and the new one."…

The problems of the streets in urban areas, as teachers often note, frequently spill over into public schools. In the public schools of East St. Louis this is literally the case.

"Martin Luther King Junior High School," notes the *Post-Dispatch* in a story published in the early spring of 1989, "was evacuated Friday afternoon after sewage flowed into the kitchen…. The kitchen was closed and students were sent home." On Monday, the paper continues, "East St. Louis Senior High School was awash in sewage for the second time this year." The school had to be shut because of "fumes and backed-up toilets." Sewage flowed into the basement, through the floor, then up into the kitchen and the students' bathrooms. The backup, we read, "occurred in the food preparation areas."

School is resumed the following morning at the high school, but a few days later the overflow recurs. This time the entire system is affected, since the meals distributed to every student in the city are prepared in the two schools that have been flooded. School is called off for all 16,500 students in the district. The sewage backup, caused by the failure of two pumping stations, forces officials at the high school to shut down the furnaces.

At Martin Luther King, the parking lot and gym are also flooded. "It's a disaster," says a legislator. "The streets are under water; gaseous fumes are being emitted from the pipes under the schools," she says, "making people ill."

In the same week, the schools announce the layoff of 280 teachers, 166 cooks and cafeteria workers, 25 teacher aides, 16 custodians and 18 painters, electricians, engineers and plumbers. The president of the teachers' union says the cuts, which will bring the size of kindergarten and primary classes up to 30 students, and the size of fourth to twelfth grade classes up to 35, will have "an unimaginable impact" on the students.

"If you have a high school teacher with five classes each day and between 150 and 175 students ... , it's going to have a devastating effect." The school system, it is also noted, has been using more than 70 "permanent substitute teachers," who are paid only $10,000 yearly, as a way of saving money....

East St. Louis, says the chairman of the state board, "is simply the worst possible place I can imagine to have a child brought up.... The community is in desperate circumstances." Sports and music, he observes, are, for many children here, "the only avenues of success." Sadly enough, no matter how it ratifies the stereotype, this is the truth; and there is a poignant aspect to the fact that, even with class size soaring and one quarter of the system's teachers being given their dismissal, the state board of education demonstrates its genuine but skewed compassion by attempting to leave sports and music untouched by the overall austerity.

Even sports facilities, however, are degrading by comparison with those found and expected at most high schools in America. The football field at East St. Louis High is missing almost everything—including goalposts. There are a couple of metal pipes—no crossbar, just the pipes. Bob Shannon, the football coach, who has to use his personal funds to purchase footballs and has had to cut and rake the football field himself, has dreams of having goalposts someday. He'd also like to let his students have new uniforms. The ones they wear are nine years old and held together somehow by a patchwork of repairs. Keeping them clean is a problem, too. The school cannot afford a washing machine. The uniforms are carted to a corner laundromat with fifteen dollars' worth of quarters....

In the wing of the school that holds vocational classes, a damp, unpleasant odor fills the halls. The school has a machine shop, which cannot be used for lack of staff, and a woodworking shop. The only shop that's occupied this morning is the auto-body class. A man with long blond hair and wearing a white sweat suit swings a paddle to get children in their chairs. "What we need the most is new equipment," he reports. "I have equipment for alignment, for example, but we don't have money to install it. We also need a better form of egress. We bring the

cars in through two other classes." Computerized equipment used in most repair shops, he reports, is far beyond the high school's budget. It looks like a very old gas station in an isolated rural town....

The science labs at East St. Louis High are 30 to 50 years outdated. John McMillan, a soft-spoken man, teaches physics at the school. He shows me his lab. The six lab stations in the room have empty holes where pipes were once attached. "It would be great if we had water," says McMillan....

Leaving the chemistry labs, I pass a double-sized classroom in which roughly 60 kids are sitting fairly still but doing nothing. "This is supervised study hall," a teacher tells me in the corridor. But when we step inside, he finds there is no teacher. "The teacher must be out today," he says.

Irl Solomon's history classes, which I visit next, have been described by journalists who cover East St. Louis as the highlight of the school. Solomon, a man of 54 whose reddish hair is turning white, has taught in urban schools for almost 30 years. A graduate of Brandeis University, he entered law school but was drawn away by a concern with civil rights. "After one semester, I decided that the law was not for me. I said, 'Go and find the toughest place there is to teach. See if you like it.' I'm still here....

"I have four girls right now in my senior home room who are pregnant or have just had babies. When I ask them why this happens, I am told, 'Well, there's no reason not to have a baby. There's not much for me in public school.' The truth is, that's a pretty honest answer. A diploma from ghetto high school doesn't count for much in the United States today. So, if this is really the last education that a person's going to get, she's probably perceptive in that statement. Ah, there's so much bitterness—unfairness—there, you know. Most of these pregnant girls are not the ones who have much self-esteem....

"Very little education in the school would be considered academic in the suburbs. Maybe 10 to 15 percent of students are in truly academic programs. Of the 55 percent who graduate, 20 percent may go to four-year colleges: something like 10 percent of any entering class. Another 10 to 20 percent may get some other kind of higher education. An equal number join the military.... "I don't go to physics class, because my lab has no equipment," says one student. "The typewriters in my typing class don't work. The women's toilets...." She makes a sour face. "I'll be honest," she says. "I just don't use the toilets. If I do, I come back into class and I feel dirty."

"I wanted to study Latin," says another student. "But we don't have Latin in this school."

"We lost our only Latin teacher," Solomon says.

A girl in a white jersey with the message DO THE RIGHT THING on the front raises her hand. "You visit other schools," she says. "Do you think the children in this school are getting what we'd get in a nice section of St. Louis?"

I note that we are in a different state and city.

"Are we citizens of East St. Louis or America?" she asks....

In a seventh grade social studies class, the only book that bears some relevance to black concerns— its title is *The American Negro*—bears a publication date of 1967. The teacher invites me to ask the class some questions. Uncertain where to start, I ask the students what they've learned about the civil rights campaigns of recent decades.

A 14-year-old girl with short black curly hair says this: "Every year in February we are told to read the same old speech of Martin Luther King. We read it every year. 'I have a dream....' It does begin to seem—what is the word?" She hesitates and then she finds the word: "perfunctory."

I ask her what she means.

"We have a school in East St. Louis named for Dr. King," she says. "The school is full of sewer water and the doors are locked with chains. Every student in that school is black. It's like a terrible joke on history."

It startles me to hear her words, but I am startled even more to think how seldom any press reporter has observed the irony of naming segregated schools for Martin Luther King. Children reach the heart of these hypocrisies much quicker than the grown-ups and the experts do....

★★★

The train ride from Grand Central Station to suburban Rye, New York, takes 35 to 40 minutes.

The high school is a short ride from the station. Built of handsome gray stone and set in a landscaped campus, it resembles a New England prep school. On a day in early June of 1990, I enter the school and am directed by a student to the office.

The principal, a relaxed, unhurried man who, unlike many urban principals, seems gratified to have me visit in his school, takes me in to see the auditorium, which, he says, was recently restored with private charitable funds ($400,000) raised by parents. The crenellated ceiling, which is white and spotless, and the polished dark-wood paneling contrast with the collapsing structure of the auditorium at [another school I visited]. The principal strikes his fist against the balcony: "They made this place extremely solid." Through a window, one can see the spreading branches of a beech tree in the central courtyard of the school.

In a student lounge, a dozen seniors are relaxing on a carpeted floor that is constructed with a number of tiers so that, as the principal explains, "they can stretch out and be comfortable while reading."

The library is wood-paneled, like the auditorium. Students, all of whom are white, are seated at private carrels, of which there are approximately 40. Some are doing homework; others are looking through the *New York Times*. Every student that I see during my visit to the school is white or Asian, though I later learn there are a number of Hispanic students and that 1 or 2 percent of students in the school are black.

According to the principal, the school has 96 computers for 546 children. The typical student, he says, studies a foreign language for four or five years, beginning in the junior high school, and a second foreign language (Latin is available) for two years. Of 140 seniors, 92 are now enrolled in AP [advanced placement] class. Maximum teacher salary will soon reach $70,000. Per pupil funding is above $12,000 at the time I visit.

The students I meet include eleventh and twelfth graders. The teacher tells me that the class is reading Robert Coles, Studs Terkel, Alice Walker. He tells me I will find them more than willing to engage me in debate, and this turns out to be correct.

Primed for my visit, it appears, they arrow in directly on the dual questions of equality and race.

Three general positions soon emerge and seem to be accepted widely. The first is that the fiscal inequalities "do matter very much" in shaping what a school can offer ("That is obvious," one student says) and that any loss of funds in Rye, as a potential consequence of future equalizing, would be damaging to many things the town regards as quite essential.

The second position is that racial integration—for example, by the busing of black children from the city or a nonwhite suburb to this school—would meet with strong resistance, and the reason would not simply be the fear that certain standards might decline. The reason, several students say straightforwardly, is "racial" or, as others say it, "out-and-out racism" on the part of adults.

The third position voiced by many students, but not all, is that equity is basically a goal to be desired and should be pursued for moral reasons, but "will probably make no major difference" since poor children "still would lack the motivation" and "would probably fail in any case because of other problems."

At this point, I ask if they can truly say "it wouldn't make a difference" since it's never been attempted. Several students then seem to rethink their views and says that "it might work, but it would have to start with preschool and the elementary grades" and "it might be 20 years before we'd see a difference."

At this stage in the discussion, several students speak with some real feeling of the present inequalities, which, they say, are "obviously unfair," and one student goes a little further and proposes that "we need to change a lot more than the schools." Another says she'd favor racial integration "by whatever means—including busing—even if the parents disapprove." But a contradictory opinion also is expressed with a good deal of fervor and is stated by one student in a rather biting voice: "I don't see why we should do it. How could it be of benefit to us?"

Throughout the discussion, whatever the views the children voice, there is a degree of unreality

about the whole exchange. The children are lucid and their language is well chosen and their arguments well made, but there is a sense that they are dealing with an issue that does not feel very vivid, and that nothing that we say about it to each other really matters since it's "just a theoretical discussion." To a certain degree, the skillfulness and cleverness that they display seem to derive precisely from this sense of unreality. Questions of unfairness feel more like a geometric problem than a matter of humanity or conscience. A few of the students do break through the note of unreality, but, when they do, they cease to be so agile in their use of words and speak more awkwardly. Ethical challenges seem to threaten their effectiveness. There is the sense that they were skating over ice and that the issues we addressed were safely frozen underneath. When they stop to look beneath the ice they start to stumble. The verbal competence they have acquired here may have been gained by building walls around some regions of the heart.

"I don't think that busing students from their ghetto to a different school would do much good," one student says. "You can take them out of the environment, but you can't take the environment out of *them*. If someone grows up in the South Bronx, he's not going to be prone to learn." His name is Max and he has short black hair and speaks with confidence. "Busing didn't work when it was tried," he says. I ask him how he knows this and he says he saw a television movie about Boston.

"I agree that it's unfair the way it is," another student says. "We have AP courses and they don't. Our classes are much smaller." But, she says, "putting them in schools like ours is not the answer. Why not put some AP classes into *their* school? Fix the roof and paint the halls so it will not be so depressing."

The students know the term "separate but equal," but seem unaware of its historical associations. "Keep them where they are but make it equal," says a girl in the front row.

A student named Jennifer, whose manner of speech is somewhat less refined and polished than that of the others, tells me that her parents came here from New York. "My family is originally from the Bronx. Schools are hell there. That's one reason that we moved. I don't think it's our responsibility to pay our taxes to provide for them. I mean, my parents used to live there and they wanted to get out. There's no point in coming to a place like this, where schools are good, and then your taxes go back to the place where you began."

I bait her a bit: "Do you mean that, now that you are not in hell, you have no feeling for the people that you left behind?"

"It has to be the people in the area who want an education. If your parents just don't care, it won't do any good to spend a lot of money. Someone else can't want a good life for you. You have got to want it for yourself." Then she adds, however, "I agree that everyone should have a chance at taking the same courses...."

I ask her if she'd think it fair to pay more taxes so that this was possible.

"I don't see how that benefits me," she says.

DISCUSSION QUESTIONS

1. Inequality permeates educational institutions, as Jonathan Kozol's essay shows. What specific inequalities in education does he note? Do you see these reflected in the school in your community?

2. How does Kozol's description of urban schools show a connection between race and class inequality?

3. One way that sociologists think about education is as a route to social mobility. Others see education as only reproducing the inequalities that exist in society. As you observe education today, what evidence do you see for each of these perspectives on the role of education in society?

INTERNET RESOURCES

Suggested Web URLs for Further Study

http://educationnorthwest.org/
This website on charter schools details attempts to reform schools and reach toward equality through charter school initiatives.

http://www.bread.org/learn/links.html
This web page provides a listing of selected non-governmental organizations that are active and involved in the field of debt and poverty reduction.

http://nces.ed.gov/
This is the official site for the National Center for Education Statistics (NCES). This site features information on literacy, education costs, student assessment, and more in the United States and around the world.

Virtual Society

For further information on this subject, including links to relevant websites, go to the Wadsworth Sociology homepage at *http://www.cengage.com/sociology*.

15

The Protestant Ethic
and the Spirit of Capitalism

MAX WEBER

A product of modern European civilization, studying any problem of univer-sal history, is bound to ask himself to what combination of circumstances the fact should be attributed that in Western civilization, and in Western civili-zation only, cultural phenomena have appeared which (as we like to think) lie in a line of development having *universal* significance and value.... All over the world there have been merchants, wholesale and retail, local and engaged in foreign trade....

But in modern times the Occident has developed, in addition to this, a very different form of capitalism which has appeared nowhere else: the rational capi-talistic organization of (formally) free labour. Only suggestions of it are found elsewhere. Even the organization of unfree labour reached a considerable degree of rationality only on plantations and to a very limited extent in the *Ergasteria* of antiquity. In the manors, manorial workshops, and domestic industries on estates with serf labour it was probably somewhat less developed. Even real domestic industries with free labour have definitely been proved to have existed in only a few isolated cases outside the Occident....

Rational industrial organization, attuned to a regular market, and neither to political nor irrationally speculative opportunities for profit, is not, however, the only peculiarity of Western capitalism. The modern rational organization of the capitalistic enterprise would not have been possible without two other impor-tant factors in its development: the separation of business from the household, which completely dominates modern economic life, and closely connected with it, rational bookkeeping....

SOURCE: From *The Protestant Ethic and the Spirit of Capitalism* by Max Weber, © 1958 Prentice-Hall, Inc. Reprinted by permission of Pearson Education, NJ: Upper Saddle River.

Hence in a universal history of culture the central problem for us is not, in the last analysis, even from a purely economic viewpoint, the development of capitalistic activity as such, differing in different cultures only in form: the adventurer type, or capitalism in trade, war, politics, or administration as sources of gain. It is rather the origin of this sober bourgeois capitalism with its rational organization of free labour. Or in terms of cultural history, the problem is that of the origin of the Western bourgeois class and of its peculiarities, a problem which is certainly closely connected with that of the origin of the capitalistic organization of labour, but is not quite the same thing. For the bourgeois as a class existed prior to the development of the peculiar modern form of capitalism, though, it is true, only in the Western hemisphere.

Now the peculiar modern Western form of capitalism has been, at first sight, strongly influenced by the development of technical possibilities. Its rationality is today essentially dependent on the calculability of the most important technical factors. But this means fundamentally that it is dependent on the peculiarities of modern science, especially the natural sciences based on mathematics and exact and rational experiment. On the other hand, the development of these sciences and of the technique resting upon them now receives important stimulation from these capitalistic interests in its practical economic application. It is true that the origin of Western science cannot be attributed to such interests. Calculation, even with decimals, and algebra have been carried on in India, where the decimal system was invented. But it was only made use of by developing capitalism in the West, while in India it led to no modern arithmetic or bookkeeping. Neither was the origin of mathematics and mechanics determined by capitalistic interests. But the *technical* utilization of scientific knowledge, so important for the living conditions of the mass of people, was certainly encouraged by economic considerations, which were extremely favourable to it in the Occident. But this encouragement was derived from the peculiarities of the social structure of the Occident. We must hence ask, from *what* parts of that structure was it derived, since not all of them have been of equal importance?

Among those of undoubted importance are the rational structures of law and of administration. For modern rational capitalism has need, not only of the technical means of production, but of a calculable legal system and of administration in terms of formal rules. Without it adventurous and speculative trading capitalism and all sorts of politically determined capitalisms are possible, but no rational enterprise under individual initiative, with fixed capital and certainty of calculations. Such a legal system and such administration have been available for economic activity in a comparative state of legal and formalistic perfection only in the Occident. We must hence inquire where that law came from. Among other circumstances, capitalistic interest have in turn undoubtedly also helped, but by no means alone nor even principally, to prepare the way for the predominance in law and administration of a class of jurists specially trained in rational law. But these interests did not themselves create that law. Quite different forces were at work in this development. And why did not the capitalistic interests do the same in China or India? Why did not the scientific, the artistic, the political, or the economic development there enter upon that path of rationalization which is peculiar to the Occident?

For in all the above cases it is a question of the specific and peculiar rationalism of Western culture.... It is hence our first concern to work out and to explain genetically the special peculiarity of Occidental rationalism, and within this field that of the modern Occidental form. Every such attempt at explanation must, recognizing the fundamental importance of the economic factor, above all take account of the economic conditions. But at the same time the opposite correlation must not be left out of consideration. For though the development of economic rationalism is partly dependent on rational technique and law, it is at the same time determined by the ability and disposition of men to adopt certain types of practical rational conduct. When these types have been obstructed by spiritual obstacles, the development of rational economic conduct has also met serious inner resistance. The magical and religious forces, and the ethical ideas of duty based upon them, have in the past always been among the most important formative influences.

on conduct. In the studies collected here we shall be concerned with these forces.

Two older essays have been placed at the beginning which attempt, at one important point, to approach the side of the problem which is generally most difficult to grasp: the influence of certain religious ideas on the development of an economic spirit, or the *ethos* of an economic system. In this case we are dealing with the connection of the spirit of modern economic life with the rational ethics of ascetic Protestantism. Thus we treat here only one side of the causal chain....

[T]hat side of English Puritanism which was derived from Calvinism gives the most consistent religious basis for the idea of the calling.... For the saints' everlasting rest is in the next world; on earth man must, to be certain of his state of grace, "do the works of him who sent him, as long as it is yet day." Not leisure and enjoyment, but only activity serves to increase the glory of God according to the definite manifestations of His will.

Waste of time is thus the first and in principle the deadliest of sins. The span of human life is infinitely short and precious to make sure of one's own election. Loss of time through sociability, idle talk, luxury, even more sleep than is necessary for health, six to at most eight hours, is worthy of absolute moral condemnation. It does not yet hold, with Franklin, that time is money, but the proposition is true in a certain spiritual sense. It is infinitely valuable because every hour lost is lost to labour for the glory of God. Thus inactive contemplation is also valueless, or even directly reprehensible if it is at the expense of one's daily work....

[T]he same prescription is given for all sexual temptation as is used against religious doubts and a sense of moral unworthiness: "Work hard in your calling." But the most important thing was that even beyond that labour came to be considered in itself the end of life, ordained as such by God. St. Paul's "He who will not work shall not eat" holds unconditionally for everyone. Unwillingness to work is symptomatic of the lack of grace.

Here the difference from the mediæval viewpoint becomes quite evident. Thomas Aquinas also gave an interpretation of that statement of St. Paul.

But for him labour is only necessary *naturali ratione* for the maintenance of individual and community. Where this end is achieved, the precept ceases to have any meaning. Moreover, it holds only for the race, not for every individual. It does not apply to anyone who can live without labour on his possessions, and of course contemplation, as a spiritual form of action in the Kingdom of God, takes precedence over the commandment in its literal sense. Moreover, for the popular theology of the time, the highest form of monastic productivity lay in the increase of the *Thesaurus ecclesliæ* through prayer and chant....

For everyone without exception God's Providence has prepared a calling, which he should profess and in which he should labour. And this calling is not, as it was for the Lutheran, a fate to which he must submit and which he must make the best of, but God's commandment to the individual to work for the divine glory. This seemingly subtle difference had far-reaching psychological consequences, and became connected with a further development of the providential interpretation of the economic order which had begun in scholasticism.

It is true that the usefulness of a calling, and thus its favour in the sight of God, is measured primarily in moral terms, and thus in terms of the importance of the goods produced in it for the community. But a further, and, above all, in practice the most important, criterion is found in private profitableness. For if that God, whose hand the Puritan sees in all the occurrences of life, shows one of His elect a chance of profit, he must do it with a purpose. Hence the faithful Christian must follow the call by taking advantage of the opportunity. "If God show you a way in which you may lawfully get more than in another way (without wrong to your soul or to any other), if you refuse this, and choose the less gainful way, you cross one of the ends of your calling, and you refuse to be God's steward, and to accept His gifts and use them for Him when He requireth it: you may labour to be rich for God, though not for the flesh and sin."...

The superior indulgence of the *seigneur* and the parvenu ostentation of the *nouveau riche* are equally detestable to asceticism. But, on the other hand, it has the highest ethical appreciation of the sober, middle-class, self-made man. "God blesseth His trade" is a stock remark about those good men who had successfully followed the divine hints. The whole power of the God of the Old Testament, who rewards His people for their obedience in this life, necessarily exercised a similar influence on the Puritan who ... compared his own state of grace with that of the heroes of the Bible....

Although we cannot here enter upon a discussion of the influence of Puritanism in all ... directions, we should call attention to the fact that the toleration of pleasure in cultural goods, which contributed to purely aesthetic or athletic enjoyment, certainly always ran up against one characteristic limitation: They must not cost anything. Man is only a trustee of the goods which have come to him through God's grace. He must, like the servant in the parable, give an account of every penny entrusted to him, and it is at least hazardous to spend any of it for a purpose which does not serve the glory of God but only one's own enjoyment. What person, who keeps his eyes open, has not met representatives of this viewpoint even in the present? The idea of a man's duty to his possessions, to which he subordinates himself as an obedient steward, or even as an acquisitive machine, bears with chilling weight on his life. The greater the possessions the heavier, if the ascetic attitude toward life stands the test, the feeling of responsibility for them, for holding them undiminished for the glory of God and increasing them by restless effort. The origin of this type of life also extends in certain roots, like so many aspects of the spirit of capitalism, back into the Middle Ages. But it was in the ethic of ascetic Protestantism that it first found a consistent ethical foundation. Its significance for the development of capitalism is obvious.

This worldly Protestant asceticism, as we may recapitulate up to this point, acted powerfully against the spontaneous enjoyment of possessions; it restricted consumption, especially of luxuries. On the other hand, it had the psychological effect of freeing the acquisition of goods from the inhibitions of traditionalistic ethics. It broke the bonds of the impulse of acquisition in that it not only

legalized it, but (in the sense discussed) looked upon it as directly willed by God....

As far as the influence of the Puritan outlook extended, under all circumstances—and this is, of course, much more important than the mere encouragement of capital accumulation—it favoured the development of a rational bourgeois economic life; it was the most important, and above all the only consistent influence in the development of that life. It stood at the cradle of the modern economic man.

To be sure, these Puritanical ideals tended to give way under excessive pressure from the temptations of wealth, as the Puritans themselves knew very well. With great regularity we find the most genuine adherents of Puritanism among the classes which were rising from a lowly status, the small bourgeois and farmers while the *beati possidentes*, even among Quakers, are often found tending to repudiate the old ideals. It was the same fate which again and again befell the predecessor of this worldly asceticism, the monastic asceticism of the Middle Ages. In the latter case, when rational economic activity had worked out its full effects by strict regulation of conduct and limitation of consumption, the wealth accumulated either succumbed directly to the nobility, as in the time before the Reformation, or monastic discipline threatened to break down, and one of the numerous reformations became necessary.

In fact the whole history of monasticism is in a certain sense the history of a continual struggle with the problem of the secularizing influence of wealth. The same is true on a grand scale of the worldly asceticism of Puritanism. The great revival of Methodism, which preceded the expansion of English industry toward the end of the eighteenth century, may well be compared with such a monastic reform. We may hence quote here a passage from John Wesley himself which might well serve as a motto for everything which has been said above. For it shows that the leaders of these ascetic movements understood the seemingly paradoxical relationships which we have here analysed perfectly well, and in the same sense that we have given them. He wrote:

I fear, wherever riches have increased, the essence of religion has decreased in the same proportion. Therefore I do not see how it is possible, in the nature of things, for any revival of true religion to continue long. For religion must necessarily produce both industry and frugality, and these cannot but produce riches. But as riches increase, so will pride, anger, and love of the world in all its branches. How then is it possible that Methodism, that is, a religion of the heart, though it flourishes now as a green bay tree, should continue in this state? For the Methodists in every place grow diligent and frugal; consequently they increase in goods. Hence they proportionately increase in pride, in anger, in the desire of the flesh, the desire of the eyes, and the pride of life. So, although the form of religion remains, the spirit is swiftly vanishing away. Is there no way to prevent this—this continual decay of pure religion? We ought not to prevent people from being diligent and frugal; *we must exhort all Christians to gain all they can, and to save all they can; that is, in effect, to grow rich.*

As Wesley here says, the full economic effect of those great religious movements, whose significance for economic development lay above all in their ascetic educative influence, generally came only after the peak of the purely religious enthusiasm was past. Then the intensity of the search for the Kingdom of God commenced gradually to pass over into sober economic virtue; the religious roots died out slowly, giving way to utilitarian worldliness. Then, as Dowden puts it, as in *Robinson Crusoe*, the isolated economic man who carries on missionary activities on the side takes the place of the lonely spiritual search for the Kingdom of Heaven of Bunyan's pilgrim, hurrying through the market-place of Vanity.…

A specifically bourgeois economic ethic had grown up. With the consciousness of standing in the fullness of God's grace and being visibly blessed by Him, the bourgeois business man, as long as he remained within the bounds of formal correctness, as long as his moral conduct was spotless and the use to which he put his wealth was not objectionable, could follow his pecuniary interests as he would and feel that he was fulfilling a duty in doing so. The power of religious asceticism provided him in addition with sober, conscientious, and unusually industrious workmen, who clung to their work as to a life purpose willed by God.

Finally, it gave him the comforting assurance that the unequal distribution of the goods of this world was a special dispensation of Divine Providence, which in these differences, as in particular grace, pursued secret ends unknown to men.…

One of the fundamental elements of the spirit of modern capitalism, and not only of that but of all modern culture: Rational conduct on the basis of the idea of the calling, was born—that is what this discussion has sought to demonstrate—from the spirit of Christian asceticism. One has only to reread the passage from Franklin, quoted at the beginning of this essay, in order to see that the essential elements of the attitude which was there called the spirit of capitalism are the same as what we have just shown to be the content of the Puritan worldly asceticism, only without the religious basis, which by Franklin's time had died away.…

Since asceticism undertook to remodel the world and to work out its ideals in the world, material goods have gained an increasing and finally an inexorable power over the lives of men as at no previous period in history. Today the spirit of religious asceticism—whether finally, who knows?—has escaped from the cage. But victorious capitalism, since it rests on mechanical foundations, needs its support no longer. The rosy blush of its laughing heir, the Enlightenment, seems also to be irretrievably fading, and the idea of duty in one's calling prowls about in our lives like the ghost of dead religious beliefs. Where the fulfilment of the calling cannot directly be related to the highest spiritual and cultural values, or when, on the other hand, it need not be felt simply as economic compulsion, the individual generally abandons the attempt to justify it at all. In the field of its highest development, in the United States, the pursuit of wealth, stripped of its religious and ethical meaning, tends to become associated with purely mundane passions, which often actually give it the character of sport.

No one knows who will live in this cage in the future, or whether at the end of this tremendous development entirely new prophets will arise, or there will be a great rebirth of old ideas and ideals, or, if neither, mechanized petrification, embellished with a sort of convulsive self-importance.…

Here we have only attempted to trace the fact and the direction of its influence to their motives in one, though a very important point. But it would also further be necessary to investigate how Protestant

Asceticism was in turn influenced in its development and its character by the totality of social conditions, especially economic. The modern man is in general, even with the best will, unable to give religious ideas a significance for culture and national character which they deserve. But it is, of course, not my aim to substitute for a one-sided materialistic an equally one-sided spiritualistic causal interpretation of culture and of history. Each is equally possible, but each, if it does not serve as the preparation, but as the conclusion of an investigation, accomplishes equally little in the interest of historical truth.

DISCUSSION QUESTIONS

1. What relationships did Max Weber find between socioeconomic status and religion? Is this what you would have expected?

2. According to Weber, what is the spirit of capitalism? Is it an ideal type? Can you think of an example of another ideal type?

3. What are the specific components of the spirit of capitalism? Do you think they still reflect

important cultural values in contemporary western society? According to whom?

4. What does Weber mean when he argues that cultural values circumscribe social action?

5. In a historical context, why did capitalism emerge in the West?

INTERNET RESOURCES

Suggested Web URLs for Further Study

http://media.pfeiffer.edu/lridener/DSS/
This is the Dead Sociologists Index.

http://www.maxweber.com
This is a site dedicated entirely to Max Weber's works and ideas; it also contains further links.

http://www.maxweberstudies.org
This is a journal dedicated entirely to spreading Weber's ideas.

Virtual Society

For further information on this subject, including links to relevant websites, go to the Wadsworth Sociology homepage at *http://www.cengage.com/sociology*.

Glossary

accommodation the norms that develop between ethnic groups regarding the ways they will interact with one another; the second stage of Park's race relations cycle

achieved status position that is attained through individual efforts, such as one's education or occupation

advanced industrial societies term used to refer to societies that have very productive extractive and manufacturing sectors and a growing service sector and a complex and productive communications industry

age-based norms expectations and rules about behaviors that should occur at various ages

age effect phenomenon whereby age-based norms influence our behaviors and actions

age/sex structure see *population structure*

agent of socialization the person or group that provides information about social roles

aggregates large groups of people who actually have no relationship to one another except that they might happen to be in the same place at the same time

agrarian societies societies with intense agricultural production made possible by the plow

alienation term, first used by Marx, that refers to the separation of workers from the product or result of their work, which can result in feelings of powerlessness

anomie Durkheim's term describing a situation of uncertainty over norms or normlessness

anticipatory socialization the process by which we prepare ourselves for future roles through thinking about

and rehearsing the actions, emotions, and skills that may be involved in these new roles

ascribed status position that is attained through circumstances of birth and that cannot be changed, such as one's race/ethnicity or gender

assimilation the merger of two ethnic groups; the final stage of Park's race relations cycle

attributes individual characteristics of persons or things; the values of variables

back-channel feedback verbal and nonverbal conversational techniques used to let others know whether their spoken messages are being understood

backstage Goffman's term for the setting, or frame, in which impression management is not needed; contrast with *front-stage*

base term used by Marx and Marxian scholars to refer to the economy and the dynamics surrounding the means of production, which he saw as the basis of social structure

birth cohort people who are born in the same year; see also *cohort*

birth rates measures of the number of births in a society relative to the total number of people in the society

body language physical movements, postures, and gestures

bounded rationality term used to refer to the ways in which decisions made within organizations are constrained, or bounded, by the structure, technology, and culture of the organization

bourgeoisie term used by Marx and Marxian scholars to describe capitalists, people who own the factories and mills; contrast with *proletariat*

bureaucracies highly structured and formalized organizations that are governed by laws and rules

capitalism the free-enterprise economic system in which private individuals or corporations develop, own, and control business enterprises; contrast with *socialism*

capitalists people who own large industries, who continually reinvest the profits from these industries to increase their wealth, and whose power and social standing come from their control of capital (money and other tangible resources)

career the sequence of roles that we enact during our lifetimes

cartel organization of nations that produce similar products and cooperate to limit competition among one another and jointly set prices for their products

case study a research technique that involves an in-depth look at one case, such as one person, one group, or one organization

caste system a system of stratification like that traditionally found in India, whereby the family into which one is born determines one's social status and the types of occupations one can hold

causal model graphic device that illustrates sociological theories involving causal relationships between variables

causal relationship an association between variables in which one influences or causes the other

census a complete count of a group of people

census tracts small, relatively permanent areas within cities and towns that are designed by the Census Bureau to be fairly homogeneous in terms of population, economic situation, and living conditions and that represent, as much as possible, a city's neighborhood boundaries

charismatic authority Weber's term for authority that is based on the personal qualities of a particular individual

church a religious group that accepts the social environment in which it exists; compare with *sect*

class consciousness Marx's term for people in a social class who are aware of their common interests and concerns and of the fact that these interests conflict with those of another class group

closed-ended questions survey questions that give respondents only certain possible options from which to choose their answers

cognitive development emerging ability to understand and interpret the world

cognitive theory of social movements a theory of social movements that focuses on individuals' ideas and cognitive understandings and the ways in which these both prompt involvement in social movements and are shaped and developed by the actions of social movements

cohabitation the practice of men and women living together without marriage

cohort group of individuals who have some type of characteristic in common, such as the year in which they were born

cohort effect the common influence on attitudes and behaviors stemming from similar experiences based on cohort membership

collective behavior actions of groups of people, some of whom may not even know each other, that are often unexpected, spontaneous, and may involve unusual or atypical behaviors

collective consciousness Durkheim's term describing the common beliefs, values, and norms of people within a society

collective identity joint agreements about the nature of the group, who belongs and who does not belong, and what the group should and should not do

command economy an economic system in which the political elite determine how goods and services will be produced, distributed, and consumed and in which workers' labor is coerced rather than freely hired

common schools schools held "in common"; that is, for all children to attend

community a group of people who have frequent face-to-face interactions and common values and interests, relatively enduring ties, and a sense of personal closeness to one another

comparative research research method that involves the comparison of data from a variety of groups or settings, such as nations or historical eras

compensating differentials theory that employers compensate their employees for working in undesirable working conditions by paying them more

competition the inevitable conflicts between new and entrenched ethnic groups; the first stage in Park's race relations cycle

computer simulation the use of computer programs to simulate, or model, selected characteristics of real life;

can be used by sociologists to conduct hypothetical experiments

concentric–zone model model of urban growth suggesting that cities expand outward from the central business district in a series of concentric circles or zones

concrete operational stage Piaget's term for the developmental stage, from about ages seven to eleven, in which children become able to handle logical operations

conglomerate corporate structure that includes many different firms specializing in entirely different types of products

conjugal family system a kinship system that emphasizes marriage ties over blood ties

consanguine family system a kinship system that emphasizes blood ties over marriage ties

conscience an internal set of ethical and moral principles that guide actions

construction industry industry that uses raw materials to build homes, offices, roads, and other structures

contact hypothesis hypothesis suggesting that individual discrimination and prejudice toward members of a minority group will diminish once members of the two groups have direct interpersonal contact

content analysis research method that involves the study of written documents or other types of recorded materials, such as pictures, movies, or television shows

contingency approach theory of organizational leadership that suggests effective leadership traits and behaviors vary depending on a particular situation

control theory theory emphasizing that one's connections to others within society are the major influence on one's desire to conform to society's norms

control variable variable that is added to an analysis to see if it affects the relationship between an independent and a dependent variable

convergence theory theory that most countries around the world are moving toward the establishment of a mixed economic system

conversation analysis research technique involving close examination of patterns in verbal interactions

core nations term used by world systems theorists to refer to the most powerful, dominant countries in the global economy

corporate actors a unified group, such as an organization or nation, that engages in social actions

correlation the ways in which two variables may be related to each other in a predictable pattern

crime actions that a society explicitly prohibits and that are sanctioned through official means

crime rate a calculation of how frequently crime occurs for every 100,000 people within a population

cult a religion that is in its beginning stages and is new to a society; the same as a new religious movement

cultural capital Bourdieu's term describing how people behave, dress, and talk and how these manners and styles differentiate those in one class group from those in another

cultural diffusion process whereby elements of one culture or subculture spread from one society or culture to another

cultural elements basic aspects or characteristics of a culture, both material and nonmaterial

cultural gatekeepers powerful organizations and individuals that control the entrance of new cultural elements into a society

cultural identity the way in which people come to see themselves as belonging to or being part of a culture or cultural group

cultural innovation process whereby new cultural elements are created

cultural lag period of delay when one part of a society changes and other parts of a society have not yet readjusted

cultural materialism theoretical perspective that suggests that societies develop differently primarily because they exist in different environments and thus tap different resources or materials

cultural universals characteristics of a society that appear in all or virtually all cultures

culture a common way of life; the complex pattern of living that humans develop and pass on from generation to generation

curvilinear a relationship between two variables that is not well-represented by a straight line and is better represented by some type of curved line

data factual information that is used as the basis for making decisions and drawing conclusions; the plural form of the word datum

deductive reasoning logical process of reasoning that moves from general theories or ideas to specific hypotheses or expectations; the opposite of inductive reasoning

deforestation the disappearance of forested land, usually caused by overcutting of trees

deindustrialization a shift in economy from producing goods to producing services

democracy government by the people, especially majority rule

democratic ideology the belief that the state belongs to the people and should serve the interests of all citizens

democratic organization an organization that is responsive to the desires of its members and does not degenerate into an oligarchy

democratic socialism a mixture of capitalism and socialism whereby the government owns and operates some aspects of the economy and private industry owns and operates others; also called welfare socialism

demographic momentum the growth of a society that occurs simply because of the people, or demographic characteristics, that are already there

demographic (population) processes the ways in which populations change in composition and size

demographic transition theory theory that summarizes the changes, or transitions, in population that societies experience as they move from the high fertility and mortality typical of pre-industrial societies to the low-fertility and mortality rates seen in highly industrialized societies

demography the scientific study of populations and their effects

dependency ratio The ratio of the dependent-age population, considered to be those younger than age 15 or older than age 64, to those of the traditional working ages (ages 15–64), used as a measure of how many people the average working person has to support

dependent variable variable that is said to be influenced or caused by another variable (the independent variable)

desertification an increase in desert land, usually caused by overgrazing

developing countries countries that are in the process of industrializing and that are in the middle of the global stratification system

deviance behaviors that violate social norms

differential association theory theory that focuses on how people learn deviant roles from their association or relationships with others engaged in deviant activities

diffuse status characteristics status characteristics that go beyond specific task-related skills and reflect more general statuses, such as occupation, education, gender, age, or race/ethnicity

discrimination differential treatment accorded to a group of people based solely on ascribed characteristics, such as race/ethnicity

distributive justice see *norm of equity*

diversification form of corporate merger that results in a corporation specializing in many different aspects of one industry

divorce rate the number of divorces in a year relative to every 1,000 existing marriages

domination form of power in which one party controls the behavior of others through sanctions; compare with *influence*

dramaturgical theory theory derived from the work of Goffman that uses the metaphor of a drama to explain how individuals play social roles and thus produce social structure

dysfunctions negative consequences of a structure of society for the whole of society

ecological worldview set of beliefs suggesting that new technology may create as many problems as it solves, that Earth's resources are limited, that the world's population has reached Earth's carrying capacity, and that population growth must be curbed

economic capital income-producing wealth

economic exchanges economic transactions; the transfer of goods and services in return for some type of valued item

economy social institution that includes all the norms, organizations, roles, and activities involved in the production, distribution, and consumption of goods and services

education social institution that includes the groups, organizations, norms, roles, and statuses associated with a society's transmission of knowledge and skills to its members

educational attainment the number of years of education a person has

elites powerful people who are able to influence the political process

elitism view that political power and influence are dominated by a small handful of people who are relatively unified and form a comparatively small, tight-knit social network

emergent norms norms that develop or emerge from a setting—rather than already being part of the well-known—and specified expectations established within a culture

empirical based on experience and observation rather than pre-existing ideas

endogamy marriage rule requiring people to select partners from within their own tribe, community, social class, or racial/ethnic or other such group

epidemiology the study of epidemics, widespread outbreaks of communicable disease; today it also includes the study of other conditions that threaten health

ethnic enclave area of a community in which there is a great deal of entrepreneurial business activity by members of a particular racial-ethnic group

ethnic group group of people who have a common geographical origin and biological heritage and who share cultural elements, such as language; traditions, values, and symbols; religious beliefs; and aspects of everyday life, such as food preferences

ethnic identity feeling of belonging specifically to an ethnic group

ethnocentrism the belief that one's own culture or way of life is superior to that of others

evaluation research research designed to examine the effectiveness of social change or intervention projects

exchange theory Homans's theory that social action is an ongoing interchange of activity between rational individuals who decide whether they will perform a given action based on its relative rewards or costs

exclusion term describing attempts to entirely remove lower-paid groups from a labor market, as, for example, through restrictive immigration laws

exogamy marriage rule requiring people to select partners from outside their own tribe, community, social class, or racial/ethnic or other such group

expectation states beliefs about what others within a group can do

expectation states theory theory that group members' beliefs about what other members can do influence group interactions and the power structure within groups

experiment research method that uses control groups and experimental groups to assess whether a causal relationship exists between an independent and a dependent variable

experimental variable see *independent variable*

expressive leadership leadership that deals with interpersonal relationships within a group or organization

extended family family that includes relatives besides parents and children; contrast with nuclear family

external area term used by world systems theorists to refer to countries that are largely excluded from the global economy and that are much poorer than other countries

external social control attempts by others to control one's behavior

extractive industries industries involved in the removal of raw materials from the environment, such as agriculture, fishing, forestry, and mining

falsification the logic that underlies the testing of hypotheses; we can never prove that a theory is true but can only say that it has either been falsified (shown to be untrue) or not yet been falsified

family development theory theoretical perspective that looks at changes in family structure to understand how family life changes and how these changes affect individuals

family of orientation the family in which one grows up

family of procreation the family in which one lives as an adult and in which one has children (or procreates)

family structure the combination of statuses within a family, particularly the number and type of statuses that are included

feminism an ideology that directly challenges gender stratification and male dominance and promotes the development of a society in which men and women have equality in all areas of life

feminist movement a social movement designed to promote the interests of women, much as the civil rights movement promoted racial/ethnic equality

fertility rates measures of the number of births in a society relative to the total number of women of child-bearing age

field experiment experiment that takes place in a real-life setting as opposed to a laboratory

field research research method in which a researcher directly observes behaviors or other phenomena in their natural settings

firm the parent company or organization of work establishments

folklore myths and stories that are passed from one generation to the next within a culture

folkways norms that govern the customary way of doing everyday things

formal operations stage Piaget's term for the final developmental stage in which children become capable of more complex and abstract thought

formal organizations groups that have been deliberately created to accomplish certain goals

formal task groups groups with specific goals, established norms, and recognized membership

frame term used by dramaturgical theorists to refer to the setting in which interaction occurs and the norms that apply to a given situation

framing the way social movements present and shape the ideas that underlie social movements

free riders individuals who benefit from the public goods of a group or society even if they do not take the time or energy to participate

front-stage Goffman's term for the setting, or frame, in which behaviors are designed to impress or influence others and in which impression management is important; contrast with backstage

function the part a structure plays in maintaining or altering the society

game stage Mead's term for the stage in which children begin to understand how different roles go together and become able to take the role of the other

Gemeinschaft Tönnies's term describing relationships that might appear in small, close-knit communities in which people are involved in social networks with relatives and longtime friends and neighbors, much like those that appear in primary groups

gender socially learned expectations and behaviors associated with members of each sex

gender-based division of labor rules about what tasks members of each sex should perform

gender discrimination differential treatment of women because of their sex

gender identity one's gut-level belief that one is a male or a female

gender roles the norms and expectations associated with being male or female

gender schema cognitive framework that is used to organize information as relevant to one sex or the other

gender segregation the restriction of members of each sex group to different statuses and roles

gender segregation of the labor force the gender-based division of labor in the occupational world or the phenomenon of men and women holding very different jobs; also called occupational gender segregation

gender socialization learning to see oneself as a male or female and learning the roles and expectations associated with that sex group

gender stratification the organization of society in a way that results in members of one sex group having more access to wealth, prestige, and power than members of the other sex group

general fertility rate the number of births per every 1,000 women of childbearing age in a given year

general strain theory this theory of deviance that expands upon Merton's original formulation and suggests that the sources of structural strain can be much broader than Merton suggested and that the response which individuals choose is influenced by their perceptions of the fairness of the situation and the nature of their social networks

generalized other Mead's term for the conception people have of the expectations and norms that others generally hold; the basis of the "me"

Gesellschaft Tönnies's term describing relationships that come about through formal organizations and economic relationships rather than kinship or friendship, similar to those that appear in secondary groups

ghetto highly segregated neighborhood populated primarily by people of one racial/ethnic heritage

Gini index a measure of income inequality within societies that can vary from zero, indicating that everyone receives an equal share of the country's income, to 1, indicating that all the income is received by just one person

global economy economic exchanges and markets that include the entire world

globalization the process by which all areas of the world are becoming interdependent and linked with one another

global stratification system differences between nations that result in some societies having more power, prestige, and property than others

goods-producing industries industries involved in either the extraction, manufacture, or construction of products

grassroots organizations social movement groups that develop "from the ground up," away from major political centers

gross national product the value of all goods and services that a nation produces; GNP

group in sociological terms, two or more people who regularly and consciously interact with each other through engaging in some common activity and having some relatively stable social relationship

group boundaries patterns of behavior that define who does and who does not belong to a given group

groupthink process by which members of a group become so oriented to maintaining the cohesiveness of the group that they ignore or suppress information that may be critical of their decisions

group threat the extent to which members of a dominant group perceive that minority groups threaten their well-being; used as an explanation of variations in prejudice from one area or time period to another

growth rate a measure of how many people are added to a population each year divided by the number of people who are already in the population

hegemony a hidden but pervasive power involving such extreme domination of social life that we seldom recognize it or question its legitimacy

hierarchy of identities ordering of identities such that some identities are more important than others

historical–comparative research research that uses historical data to compare two or more societies

historical research research method that involves the examination of data from the past, often written artifacts and records

horizontal differentiation the extent to which the work of an organization is divided among different units or subgroups; part of organizational complexity

horticultural societies societies that are based on a gardening economy

human capital resources that we have as individuals and workers, such as our education, skills, and work experience

human capital theory theory that suggests that different occupations and incomes reflect different amounts of human capital

hunting and gathering societies societies that obtain their food through hunting animals and gathering plants and that, compared with other types of societies, use the simplest, most primitive tools and work techniques

hypersegregation extreme isolation from other areas of the community

hypotheses statements about the expected relationship between two or more variables; often derived from theories

ideal types concepts or descriptions of phenomena that may not exist in a pure form in the real world but that define basic aspects of a given situation

ideologies complex and involved cultural belief systems

imperialism practice whereby one powerful country forcibly acquires territories in other areas of the world, creating vast colonial empires

impression management Goffman's term describing how individuals may manipulate the impression or view that others have of them and give out cues to guide interactions in a particular direction

income how much money a person receives in a given time, such as $20,000 a year

independent variable variable that is said to cause or influence another variable (the dependent variable); also called the *experimental variable* in experiments

index crimes set of eight serious crimes used in the Uniform Crime Reports as the basis to calculate crime rates; see also *property crimes* and *violent crimes*

index of dissimilarity a measure of segregation that is calculated from data on the characteristics of people living in different census tracts; also sometimes called a segregation index

individual discrimination actions of individuals or small groups involving both discrimination and prejudice

inductive reasoning logical process of reasoning that moves from specific ideas and observations to more general hypotheses and theories, the opposite of deductive reasoning

industrialized countries countries that have a highly industrialized economy and that are at the top of the global stratification system

industrial worldview the set of beliefs involving the notion that Earth's resources are valuable because they provide for humans' needs

industry branches or areas of economic activity

infant mortality rate the number of babies in 1,000 who are expected to die before the age of one in a given year

influence form of power whereby providing information and knowledge leads others to take different actions; compare with domination

informal groups groups with no specified goals or formalized norms and no set membership

informal structure social networks within an organization that exist alongside the organization's formal structure

institutional norms norms that prescribe appropriate structures and behaviors of organizations and other aspects of social institutions

instrumental leadership leadership that deals with the tasks or work of an organization

interaction effect a pattern of influence in which the influence of one variable changes depending on the status of another variable

interest groups political organizations that concentrate their activities on specific policy issues or concerns

intergenerational mobility holding a different occupational or social class position than one's parents

intergenerational succession holding the same occupational or social class position as one's parents

internal migration migration within national borders

international migration migration across national borders

internalization acceptance of the norms and behaviors of the group; the development of a conscience

internal labor market the series of jobs that people may hold within an organization throughout their work careers

internal social control control over one's behavior that is based on internalized standards; self-control

intervening variable variable that comes between a dependent variable and an independent variable in a causal relationship

kin group of relatives beyond the nuclear family

labeling theory theory that suggests that definitions of deviant behavior develop from social interactions and that the key element in becoming deviant is how others respond to people's behavior, rather than how they actually behave

labor force people employed within a society and, sometimes, those actively seeking work; workforce

labor market set of jobs within a society

labor market segmentation theory theory that the labor market is divided into primary and secondary sectors and that women and members of racial/ethnic minorities earn less than others because they more often hold jobs in the secondary sector

labor unions organizations of workers who have joined together to collectively promote their interests as workers

laissez-faire capitalism capitalistic system in which the government has no control or influence over the operation of business

land reform a policy that calls for redistributing very large parcels of land, many of which are now owned by extremely wealthy landowners or multinational corporations, to landless peasants

latent functions functions that are less obvious and often unintended and that are generally unnoted by the people involved; contrast with manifest functions

laws norms that are codified or written down; may be either folkways or mores

least developed countries countries that are the least industrialized and are at the bottom of the global stratification system

legal authority Weber's term for authority based on written rules

life expectancy the number of years a person is expected to live

literacy the ability to read and write

longitudinal study study that involves data that have been collected at different times

macrolevel theories and analyses that deal with relatively broad areas of society rather than with individuals

mainstreaming the practice of integrating special education students into the general curriculum and school day

male dominance cultural beliefs that give greater value and prestige to men and to their roles and activities

manifest functions functions that are easily seen and obvious; contrast with latent functions

manufacturing industries industries that process raw materials into more usable forms

market process of buying and selling; the way in which values are established for goods and services

mass schooling elementary-level schooling for all children; common schools

material culture the physical objects that are distinctive to a group of people, such as their food, clothes, houses, and hairstyles

mean statistical average; computed by simply adding up all values and dividing by the number of cases

means of production term used by Marx and Marxian scholars to refer to the way in which people produce their living, such as by farming or manufacturing or hunting and gathering

measure the way in which concepts involved in a theory are translated into actual data

mechanical solidarity Durkheim's term describing a society in which there is little differentiation, with people performing similar tasks, sharing similar responsibilities, and having similar behaviors; contrast with organic solidarity

median the midway point in a distribution; the point at which 50 percent of the cases are larger and 50 percent are smaller

megalopolis "great city"; a term used to designate a vast area of land with a dense population

megamergers form of corporate merger common since the 1980s that involves the union of some of the largest companies in the economy

mesolevel theories and analyses that deal with social groups and organizations, such as individual families, classrooms, and work groups, rather than with very broad areas of society or with individuals

meta-analyses the review and summary of a large number of studies of the same phenomenon to see what patterns emerge

methodology the rules and procedures that guide research and help make it valid

metropolitan area (MA) a region designated by the Census Bureau for statistical purposes that usually includes both the core area of a city with a large population and nearby communities

microlevel theories and analyses that deal with relatively narrow aspects of social life, such as individuals' day-to-day activities and relations with other people

middle-range theories theories that focus on relatively limited areas of the social world, as opposed to grand theories; often incorporate aspects of grand theories but are much more directed and applied toward specific research problems and can thus be more easily tested

migration rates measures of the number of people who have moved into or out of an area relative to its total population

minority group subculture that is subordinate to another group or groups within the society and that has less power, privilege, wealth, or prestige

mixed economic system an economic system that blends elements of socialism and capitalism

mobility table table of data used to illustrate patterns of intergenerational mobility and succession, typically involving the relationship between the occupations of parents and their offspring

modal the most common or frequently occurring category or case

modernization theory perspective suggesting that the majority of less-developed countries will eventually industrialize in the manner of countries such as England and the United States

monogamy being married to one person at a time

monopoly situation in which one company controls an entire market or industry

mores norms that are vital to society and the violation of which is seen as morally offensive

mortality rates measures of the number of deaths in a society relative to certain characteristics of the population

multinational corporations business corporations that have outlets around the globe

multiple-nuclei model model of urban growth suggesting that cities may have a number of different core areas (or nuclei), often influenced by the geographic and physical characteristics of an area

nation-state political entity that unites large groups of people under agreed-upon laws and regulations

natural increase the excess of births over deaths; the change in population size that comes from the difference in the number of births and deaths

neo-imperialism system whereby one country dominates another through economic means rather than military or political control, said to characterize the post–World War II relationships of the United States with other countries

neo-Marxist term used to describe recently developed theories that are in the Marxian tradition although they may depart from Marxist thought in certain ways

new religious movement a religion that is in its beginning stages and is new to a society; the same as a cult

new social movements recently developed social movements that tend to use tactics that have rarely been used before, especially the direct involvement of large numbers of citizens in political protest, and to address issues that have not been the focus of recent political debate, such as the quality of the environment, gender equality, alternative lifestyles, and human rights in developing countries

nodes see *social units*

nonmaterial culture the way of thinking of a group, including norms, values, ideology, folklore, and language

nonparticipant observation type of field research in which a researcher studies a group through observations without actually participating

nonverbal communication all the ways in which we send messages to others without words, including posture and movements, facial expressions, clothes and hairstyles, and manners of speaking

normative life stages periods during life when people are expected to perform certain activities

norm of equity belief that things should be "fair"; that we and others should receive rewards that equal what we contribute to a relationship or interaction; also called *distributive justice*

norms cultural rules defining behavior that is expected or required within a group or situation; includes folkways, mores, and laws

nuclear family family group consisting of a mother, a father, and their children; contrast with extended family

occupational gender segregation see *gender segregation of the labor force*

occupational prestige scores scores given to occupations that indicate their relative prestige ranking

occupational socialization the process of learning and identifying with the norms and roles associated with a particular occupation

occupational status the occupation that one holds; often used in reference to the relative importance of one occupation over another

occupational structure occupations available within a society in a given time period

open-ended questions survey questions that allow respondents to give whatever responses they desire

organic solidarity Durkheim's term describing a society in which tasks, responsibilities, and behaviors are more highly differentiated so people are more dependent on one another; contrast with *mechanical solidarity*

organizational centralization the extent to which organizational power is centralized and decisions are made hierarchically

organizational complexity the extent to which the work of an organization is broken up and differentiated among various units; may be horizontal, vertical, and/or spatial

organizational culture the way of life, or culture, of an organization

organizational ecology theoretical perspective that looks at the relationship between organizations and their environments

organizational formalization the extent to which an organization uses written rules and procedures to control individuals within it

organizational leaders people within organizations who are able to influence the things others do and believe

organizational structure the way that the various parts of an organization are arranged; typically includes the elements of complexity, formalization, and centralization

organizational technology the work of organizations, including the skills and knowledge of workers in an organization as well as the characteristics of the materials they work with and the machines they use

panel study a longitudinal study that includes information on the same people over a long period of time

paradigm a fundamental model or scheme that guides people's thinking about a particular subject

paradigm crisis the breakdown of an explanatory scheme, requiring a fundamental reorientation or rethinking of basic assumptions

paralanguage nonverbal aspects of speech, such as tone of voice and emphasis on words

participant observation type of field research in which a researcher studies a group or event while actually participating in it

peer review process by which research findings are evaluated by other experts in the field before publication to see if they meet the standard rules of research methodology

people of color people whose ancestors or who themselves came from non-European areas of the world and who can be identified by skin color

per capita GNP the gross national product adjusted for the number of people who live in the country

period effect the influence of the norms and events of a particular historical time on individuals' lives

peripheral nations term used by world systems theorists to refer to countries that sell raw materials to core nations and that are at the edge of the global economy

play stage Mead's term for the developmental stage in which children "play at" or assume various roles one at a time

pluralism view that the political power structure involves a number of powerful groups and individuals, all of which can potentially influence the decision-making process

plutocracy government by the wealthy

political economy the interrelationship of economic and political processes

polity the social institution that includes all the various ways that societies have developed to maintain social order and control; the political world

polyandry a system of marriage whereby the wife can have more than one husband; the opposite of polygyny

polygyny a system of plural marriage whereby the husband has more than one wife

population the entire group or set of cases that a researcher is interested in generalizing to; see also *sample*

population projections estimates of the size and composition of a population in future years

population pyramid diagram used by demographers to illustrate the population or age/sex structure of a society

population structure the age and sex composition of a society; the age/sex structure

poverty line the income level under which families are officially defined as being poor

poverty rate the percentage of the total population that lives in families with incomes under the official poverty level

power the ability of one social element, either a group or a person, to compel another social element to do what it wants

power structure the power differences within a group; the way that power is distributed in a society

PPE spiral term used to refer to the intertwining influences of poverty, population growth, and environmental degradation

prejudice preconceived hostile attitudes toward a group of people simply on the basis of their group membership

preoperational stage Piaget's term for the developmental stage, from about eighteen months to seven years of age, in which children learn to use symbols and communicate with language but do not yet have the mental flexibility to perform mental operations that involve complex relationships

prestige classes aggregates or clusters of people who possess similar characteristics, who are perceived by their fellow citizens as being similar, and who are accorded similar levels of respect and esteem

primary groups groups that include only a few people and are characterized by intimate, face-to-face interaction

primary labor market "good" jobs with good wages and benefits, comfortable working conditions, job security, and chances for advancement; contrast with secondary labor market

probability sample sample that can be generalized to a larger group, typically chosen through some type of random selection process

proletariat term used by Marx and Marxian scholars to describe the workers; contrast with bourgeoisie

property crimes crimes that do not involve the use of violence or force against individual people, such as burglary, larceny/theft, motor vehicle theft, and arson

public goods necessities of group life that individuals cannot provide by themselves but must obtain through cooperation with others; common goods

qualitative data measures of data that cannot be assigned real numbers; contrast with quantitative data

quantitative data measures that may be assigned real, or meaningful, numbers—for example, income or age

racial/ethnic group subculture that can be distinguished on the basis of skin color and ethnic heritage

racial/ethnic stratification the organization of society such that people in some racial/ethnic groups have more property, power, or prestige than do people in other groups

random selection process that gives each member of a population an equal chance of being included in a sample

rational choice theory theory suggesting that people make decisions by balancing the costs and benefits involved and choosing the actions that provide the lowest costs and the greatest benefits

rationalization a type of action that is in agreement with some process of external systematization

reference group group of people that we look to, or use for reference, in evaluating ourselves and our position in life

reflexive behavior the process of being able to think about one's own actions and mentally assume the role of other people

regional stratification differences within a country that result in some regions having more power, prestige, and property than others

relative deprivation a gap between people's expectations and the actual state of their lives

reliability the extent to which a measure yields the same results when used by different researchers on the same subject at different times

religion the social institution that deals with the area of life people regard as holy or sacred; it involves the statuses, roles, organizations, norms, and beliefs that are related to humans' relationships with the supernatural, including shared beliefs, ethical rules, rituals and ceremonies, and communities of people with common beliefs and standards

religious economy the various religious options or choices that are available within a society

replacement rate the fertility rate that is needed for a population to replace itself or to remain stable in size (net of migration)

replication repetition of an earlier study to see if the same results occur and if they hold in other settings

representative democracy a political system in which elected officials represent the citizens and make important governmental decisions

repression psychoanalytic term for the notion that people repress, or push out of conscious awareness, ideas that are uncomfortable or painful to think about

resource mobilization theory theory suggesting that social movements are successful to the extent that they can effectively mobilize or activate the resources needed to accomplish their goals

rite of passage ceremony or ritual that helps individuals deal with role transitions through the life cycle

rituals cultural ceremonies that often mark important life events

role conflict situation in which a person holds roles with incompatible norms or obligations

role theory perspective that social structure is created and maintained because people generally act in ways that conform to social roles

role transition moves, or transitions, from one role to another during the life cycle

rural development industrialization in rural areas

sample subset of a larger group or population; see also *probability sample*

sanctions social reactions to an individual's behavior, generally reflecting attempts to control the behavior; rewards and punishments

scheme a mental framework or set of rules to understand how the world works

schooling formal instruction by trained teachers, usually within schools

secondary analysis analysis of data that have already been gathered by other researchers

secondary deviance deviance that results from the process of being labeled as a deviant

secondary groups groups that involve more than just a few people and relatively distant social relations

secondary labor market "bad" jobs with low pay, few if any benefits, little job security, and few chances for advancement; contrast with primary labor market

sect a religious group that rejects the social environment in which it exists; compare with church

sector model model of urban growth suggesting that cities are often composed of sectors or segments that extend out from the central business district

secularization process of transformation to an outlook on life based on science and reason rather than on faith and supernatural explanations

segmented occupational structure division of occupations into "good jobs," characterized by stable employment, benefits, and higher salaries, and "not-so-good jobs," which lack these characteristics

segregation the division or separation of neighborhoods in ways that lead to the inclusion of some groups and the exclusion of others

segregation index see *index of dissimilarity*

self one's view of oneself as a distinct person with a clear identity

self-concept the thoughts and feelings we have about ourselves

self-identity a set of categories used to define the self; the way we think about ourselves

semiperiphery term used by world systems theorists to refer to nations that fall between the core and periphery

sensorimotor stage Piaget's term for the developmental stage, from birth to about eighteen months, in which infants learn through their senses and their movements

serial monogamy a pattern of marriage in which a person has several spouses over a lifetime but only one at a time

service industries global term used to describe the entire set of non-goods-producing industries; also used to refer to a specific set of industries that provide various nonmaterial things people want or need, such as medical care, education, social welfare, and entertainment

sex used to refer to biological identity as male or female

significant others people with whom you interact and who are emotionally important to you

situated self the subset of self-concepts or self-identities that apply to one's self-view and behaviors in a given situation

social action day-to-day decisions and actions of individuals within the social world; social actions both influence and are patterned and influenced by social structure

social capital resources or benefits people gain from their social networks

social change the way in which societies and cultures are altered over time

social class groups of people who occupy a similar level in the stratification system

social control efforts to help ensure conformity to norms

social distance the types of social ties or interactions people are willing to establish with others

social ecology the study of the ways in which people interact with their physical and geographic environment

social fact can be empirically studied, are explained by other social facts, and are external and coercive of the individual

social institutions the complex sets of statuses, roles, organizations, norms, and beliefs that meet people's basic needs within a society

socialism an economic system that involves public rather than private ownership of the means of production; contrast with capitalism

socialization the way in which we develop, through interactions with others, the ability to relate to other people and to play a part in society

social learning perspective theory of socialization suggesting that we learn behaviors by being rewarded for those that conform to norms and by imitating those that we think conform to roles we want to fill

social mobility movement between social class groups

social movement adherents organizations and individuals that believe in the goals of a social movement

social movement constituents organizations and individuals that provide resources for a social movement

social movement industry the set of social movement organizations that are all working within the same social movement

social movement organizations formal organizations that work toward the goals of a social movement

social movements organized and concerted efforts to promote social change by groups of individuals

social movement sector the entire range of activities that are aimed at changing the social structure

social network patterns or webs of social relationships; the linkages between individuals formed by social interactions

social role expectations, obligations, and norms that are associated with a particular position in a social network

social solidarity how society is held together and how members see themselves in relation to the whole society

social status positions that individuals occupy within the social structure

social stratification the organization of society in a way that results in some people having more and some people having less; divisions in a society based on social class

social structure relatively stable patterns that underlie social life; the ways in which people and groups are related to each other and the characteristics of groups that influence our behavior

social ties relationships between units of a social network

social units elements of a social network; also called *nodes*

society a group of people who live within a bounded territory and who share a common way of life

sociological imagination Mills's term describing the ability to discern patterns in social events and view personal experiences in light of these patterns

sociology the science of society; the scientific study of the social world and social interactions

spatial dispersion the extent to which the various units of an organization are spread out in different locations; an element of organizational complexity

specific status characteristics status characteristics that are closely related to the given tasks of a group; contrast with *diffuse status characteristics*

split labor market the division of the labor market into three class groups: business owners and employers, higher-paid workers, and lower-paid workers

spurious correlation correlation between two variables that only occurs because of the influence of a third variable

state the organized monopoly or control of the use of force in a society; synonymous with government or the polity

statistical discrimination theory that employers have beliefs, or stereotypes, about the relative stability and productivity of male and female workers, which they rely on in hiring employees; when these average characteristics do not apply to individual women, statistical discrimination is said to have occurred

status term used in the Weberian tradition to designate one dimension of stratification—that involving communities or social networks of people with similar lifestyles and viewpoints; synonymous with prestige

status attainment model theoretical model that describes the way variables, such as family background, individual motivations and ability, and interactions with significant others, influence the educational levels and occupations and incomes we have in adulthood

status characteristics the social statuses that people hold and the evaluations and beliefs (characteristics) that are attached to these statuses

status generalization the process by which diffuse status characteristics influence group interactions

strain theory Merton's theory that deviant behavior results when individuals accept culturally defined goals but do not have the institutionalized means of attaining these goals

stratification system see *social stratification*

structural discrimination discrimination that results from the normal and usual functioning of the society (the social structure) rather than from prejudice or from laws and norms that promote segregation or exclusion

structural functionalism sociological theory that tries to account for the nature of social order and the relationship between different parts of society by noting the ways in which these parts or structures function to maintain the entire society

structural mobility intergenerational mobility resulting from changes in the occupational structure

subcultural theory of deviance theory suggesting that deviant behavior can reflect conformity to the norms of subcultures that support deviant lifestyles

subculture group or culture that exists within a broader culture

suburb a city or town outside a city's boundaries but either adjacent to the city or within commuting distance

superstructure term used by Marx and Marxian scholars to refer to the areas of social life that they believe are influenced by the economy, such as religious beliefs, family relations, and political life

survey research method of data gathering that involves asking people questions, through either interviews or written questionnaires

sustainable development economic development that does not deplete or degrade natural resources.

symbolic interactionism theory that social interaction involves a constant process of presenting and interpreting symbols through thinking about what another person is trying to communicate through the use of symbols

symbols anything that people use to represent something else; for example, language uses the symbols of words to represent objects and ideas

synnomie the state of a society that has a strong collective consciousness and a high degree of cohesion; the opposite of *anomie*

target of socialization the person who is learning a social role; one who is being socialized

technological innovations the development of new material elements of culture that help people deal with the problems of day-to-day life

technology the application of knowledge to solve problems

theories broad systems of ideas that help explain patterns in the social world

traditional authority Weber's term for authority that is accepted because it is an integral part of the social structure and it is impossible to conceive of any other way of being

trend analysis study that examines data collected at different time points but that uses a different sample each time

turn taking term used by conversation analysts to describe the conversation process whereby one person talks and then the other talks; results from both verbal and nonverbal cues

typology a classification of a group or phenomenon into discrete categories

unconscious psychoanalytic term for thoughts and impulses individuals may have that they are not consciously aware of

unobtrusive research research method in which a researcher obtains data without directly talking to or watching people

upward mobility social mobility that involves movement to a social class position that is higher than one's parents occupied

urbanization the process of societal change that involves the movement of people from rural areas or small towns to metropolitan areas

validity the extent to which a measure actually represents the concept it is said to be measuring; when applied to a research design, indicates that we can trust the conclusions

values general standards about what is important to a group

variables logical groupings of attributes; literally, things that vary or have more than one value

vertical differentiation the number of supervisory levels in an organization; part of organizational complexity

vertical integration form of corporate merger producing companies that control the entire extraction, manufacturing, and distribution process

violent crimes crimes that involve the actual use or threat of force against people, including murder, forcible rape, robbery, and aggravated assault

vital statistics data that record the "vital events" of life, such as births, deaths, marriages, divorces, and migration

waves of data collection times of data collection in a panel study

wealth assets resulting from the accumulation of income, such as houses, cars, real estate, and stocks and bonds

welfare capitalism form of capitalism in which a broad system of laws protects the welfare of workers and consumers in their economic transactions

welfare socialism see democratic socialism

white-collar crime nonviolent crimes that generally involve fraud and deception and are committed in the workplace

white ethnics people who identify themselves as belonging to a group with European origins, usually other than Great Britain

wholesale and retail trade industry industry involving the sale of goods to stores and directly to individual consumers

work establishment the actual place where someone works; compare with firm

workforce see *labor force*

world systems theory perspective developed by Wallerstein to explain global stratification; suggests that a set of core, highly developed nations maintain their wealth by dominating and exploiting other nations, said to be on the periphery of the world economy

zero population growth the idea that the world population will stop growing when birth rates equal death rates and when, on average, each woman has only two children